FUTURE PROSPECT OF CHEMISTRY

FUTURE PROSPECT OF CHEMISTRY

By
Dr. Shardendu Kislaya

DISCOVERY PUBLISHING HOUSE PVT. LTD.
NEW DELHI-110 002

Published by:
Tilak Wasan
DISCOVERY PUBLISHING HOUSE PVT. LTD.
4831/24, Ansari Road, Prahlad Street
Darya Ganj, New Delhi-110002 (India)
Phone: +91-11-23279245, 43764432
Fax: +91-11-23253475
E-mail: parul.wasan@gmail.com
info@discoverypublishinggroup.com
discoverypublishinghouse@gmail.com
web: www.discoverypublishinggroup.com

***First Edition:* 2011**
ISBN: 978-81-8356-769-5

Future Prospect of Chemistry

Printed at:
Mehra Offset Press
Delhi

Preface

The present book deals in general properties of matter and sound. Reason fore this change is that paper division in the examination is different Universities. One of salient course syllabi features of the present edition of the book is the inclusion of a large number of typical worked out problems which will elucidate various abstract principles and theories discussed in the text. It covers the complete syllabus of physics prescribed by Technical Universities. The treatment given is simple, lucid and comprehensive. The language used is simple.

The classification of the questions, specially the short-answer type questions, will, I hope, help the students to prepare for the final examination. I have spared no pains in applying my long experience of degree class teaching to make the book useful to the students. The author in general hope. That the present book will be warmly received by the students and teachers. We shall indeed be very thankful to our colleague for their recommendations this book of for their students.

— *Author*

Contents

1

Variable of the Covalency Theory

INTRODUCTION

The covalency of an element is ordinarily equal to the number of unpaired *s* or *p* electrons present in the ground state of the atom. For example, fluorine with configuration $1s^1\ 2s^2\ 3p_x^{\ 2}\ p_y^{\ 2}\ p_z^{\ 1}$ has only one unpaired electron and so its covalency is one.

Nitrogen with configuration $1s^2\ 2s^2\ p_x^{\ 1}\ p_y^{\ 1}\ p_z^{\ 1}$ has three unpaired electrons and its covalency is three. Oxygen has two unpaired p electrons ($1s^2\ 2s_2\ 2p_z^{\ 2}\ p_y^{\ 1}\ p_z^{\ 1}$) and its covalency is 2. All these atoms do not have d orbitals in their configurations.

Sulphur atom in the ground state $3s^2\ 3p_x^{\ 2}\ p_y^{\ 1}\ p_z^{\ 1}$ like oxygen having two unpaired electrons shows covalency of two.

But sulphur atom has empty d-orbital and when it combines with more electronegative elements like fluorine, oxygen and chlorine, the energy released in the formation of electron pair bonds causes the unpairing of *s* and p-electrons which are promoted to the vacant d-orbitals.

DATIVE BOND OR SEMIPOLAR BOND

It is a special type of covalent bond in which both of the shared electrons are contributed by one and the same atom only and not by each atom as in a covalent bond. In other words, *a coordinate bond is established between two such atoms, one of which has a complete octet and possesses a pair of valency electrons while the other is short of a pair of electrons.* The atom which contributes electron pair for sharing is called the *donor* while the other atom which accepts it is called the *acceptor*. The coordinate bond is generally represented by an: arrow head (→) starting from the donor and ending at the acceptor. This mode of representation is due to Sidgwick.

$$\underset{\substack{\text{Donor}\\\text{(Complete Octe)}}}{:\ddot{A}:} + \underset{\substack{\text{Acceptor}\\\text{(Incomplete}\\\text{Octe)}}}{:B:} \longrightarrow :\ddot{A}:\ddot{B}: \text{ or } A \longrightarrow B$$

In terms of *orbital theory,* dative bond results in from overlapping of an orbital of an atom having two electrons (lone pair) by an empty orbital of the atom.

It may be clearly noted that the co-ordinate bond once formed is indistinguishable from a covalent bond.

The simplest case involving the formation of co-ordinate bond is that of oxidation of trimethylamine $(CH_3)_3N$, to trimethylamine oxide, $(CH_3)_3NO$. In trimethylamine, as in ammonia, the nitrogen atom has a lone pair of electrons. The oxygen atom, on the other hand, stands in need of two electrons to complete its octet. Hence, the combination is represented as follows.

$$\underset{\text{Donor}}{H_3C:\overset{\overset{CH_3}{}}{\underset{\underset{CH_3}{}}{\ddot{N}}}:} + \ddot{O}: \longrightarrow \underset{\text{Acceptor}}{H_3C:\overset{CH_3}{\underset{CH_3}{N}}:\ddot{O}:} \text{ or } H_3C-\overset{\overset{CH_3}{|}}{\underset{\underset{CH_3}{|}}{N}} \longrightarrow O$$

Lowry, who first proposed the above electronic formula from the amine oxide, is of the opinion that the molecule has developed polarity. The nitrogen atom has acquired a positive charge because its electrons are moving away from it while the oxygen atom acquires a negative charge because it has additional electrons. Hence, the two parts of the molecule will be held together not only by the sharing of a pair of electrons, as in simple covalent bond, but also by their mutual electrical attraction. For this reason. Lowry called the co-ordinate linkage a *semipolar bond* and he represented the structure of trimethylamine oxide as $(CH_3)_3\overset{+}{N} - \overset{-}{O}$.

A few more examples involving co-ordinate bonds are illustrated below :

(a) Combination of ammonia and boron chloride : Although, the nitrogen has completed its octet in ammonia, it has still a lone pair of electron which it can donate to boron atom (of boron trichloride) which is short of two electrons so as to complete its outermost octet.

(b) Combination of ammonia and boron trifluoride : In terms of dot representation, in ammonia, NH_3 nitrogen has a completed octet while B in BF_3 is short of two electrons. The N becomes a donor and B acts as an acceptor. The *lone pair* on N is the sharing electron pair between N and B.

$$H:\overset{H}{\underset{H}{\ddot{N}}}: + :\overset{F}{\underset{F}{\ddot{B}}}: \longrightarrow H:\overset{H}{\underset{H}{\ddot{N}}}:\overset{F}{\underset{F}{\ddot{B}}}:F \text{ or } \left\{ H-\overset{H}{\underset{H}{\overset{|}{\underset{|}{N}}}} \rightarrow \overset{F}{\underset{F}{\overset{|}{\underset{|}{B}}}} \right\}$$

(c) Combination of ammonia and aluminium chloride : Similarly, a co-ordinate bond is formed between ammonia and aluminium chloride.

$$H:\overset{H}{\underset{H}{\ddot{N}}}: + H:\overset{Cl}{\underset{Cl}{\ddot{Al}}}:Cl \longrightarrow H:\overset{H}{\underset{H}{N}}:\overset{Cl}{\underset{Cl}{Al}}:Cl \left(H-\overset{H}{\underset{H}{\overset{|}{\underset{|}{N}}}} \rightarrow \overset{Cl}{\underset{Cl}{\overset{|}{\underset{|}{Al}}}}-Cl \right)$$

(d) Formation of hydronium ion : In water molecule, oxygen possesses two lone (unshared) pairs of electrons. One of these can be used for further bond formation with species having empty orbitals like H^+ as given below.

$$H:\ddot{\underset{..}{O}}:H + H+ \longrightarrow \left(H:\ddot{\underset{H}{O}}:H \right)^+$$

(e) Formation of NH_4^+ : In ammonia, nitrogen has completed its octet while the hydrogen (H+) having no electrons accepts the lone pair of electrons from nitrogen atom in ammonia and thus acquires the helium configuration.

$$H\overset{x}{}\overset{H}{\underset{H}{\overset{x}{\underset{x}{N}}}}: + H+ \longrightarrow \left(H\overset{x}{}\overset{H}{\underset{H}{\overset{x}{\underset{x}{N}}}}:H \right)^+ \text{ or } \left(H-\overset{H}{\underset{H}{\overset{|}{\underset{|}{N}}}} \rightarrow H \right)^+$$

(f) Formation of SO_2– : In sulphur dioxide, one oxygen atom is attached to the sulphur atom by a double covalent bond and the other by a co-ordinate bond.

$$\overset{x}{_x}\ddot{O}\overset{x}{_x} :\underset{..}{S}: \underset{xx}{\ddot{O}}\overset{x}{_x} \text{ or } O = S \longrightarrow 0$$

(g) Formation of H_2O_2 : Oxygen in water has two lone pairs of electrons, out of which one pair is donated to another oxygen atom forming hydrogen peroxide.

$$\mathrm{H} \overset{\times\times}{\underset{\times\times}{\cdot \mathrm{O} \times}} \mathrm{H} \text{ or } \mathrm{H}-\underset{\downarrow}{\mathrm{O}}-\mathrm{H}$$
$$\quad \mathrm{O} \qquad\qquad\qquad \mathrm{O}$$

Characteristics of compounds containing co-ordinate bond : The main characteristics are :

(a) **Melting and boiling points** : As the co-ordinate linkage is regarded as a combination of electrovalency and covalency, the compounds containing co-ordinate linkage have, therefore, melting and boiling points higher than those of covalent compounds but lower than electrovalent compounds.

(b) **Solubility :** These compounds are sparingly soluble in associated solvents like water but they are readily soluble in organic solvents.

(c) **Stability :** The compounds containing co-ordinate linkages may be divided into two classes, first those in which the donor and acceptor are capable of independent existence, and second, and whose separate existences are not known. The former group breaks their co-ordinate linkage readily while in the second the bond is an strong as an ordinary covalent bond. An example of a substance belonging to the first group is the compound between ether and a Grignard's reagent (I), and an example of the second group is the amine oxide (II).

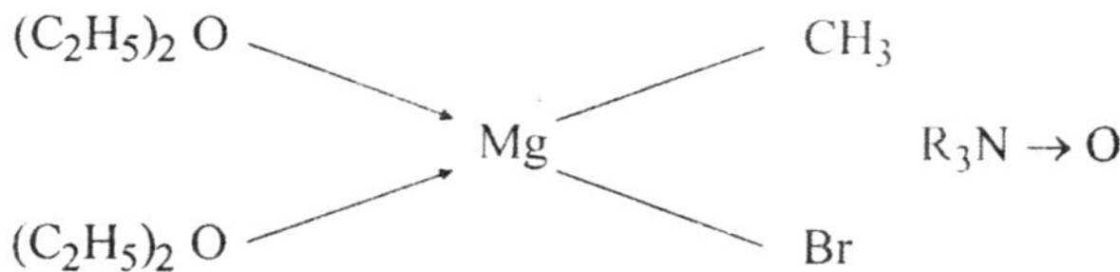

(d) **Dielectric constants** : The compounds containing co-ordinate linkage are characterised by high dielectric constant. For example, we may state the values for nitrobenzene, 36, and nitromethane, 39, as compared with 10 or less for most of the covalent compounds. Thus, regarding this property, compounds containing co-ordinate linkage have similarity with electrovalent compounds.

(e) **Stereoisomerism** : Stereochemically, the co-ordinate linkage acts as a single bond operating in a definite direction, and not as a double bond. Thus, the co-ordinate linkage is rigid and directional and thus compounds containing such a linkage show space isomerism. Regarding this property, the co-ordinate linkage is equivalent to a pure covalent bond.

(f) **Conductivity :** Compounds containing co-ordinate linkage, due to their partial covalent nature, do not ionise and are thus poor conductors of electricity

This results in the unpairing of electrons and therefore increases the covalency of sulphur.

Oxygen atom in ground state

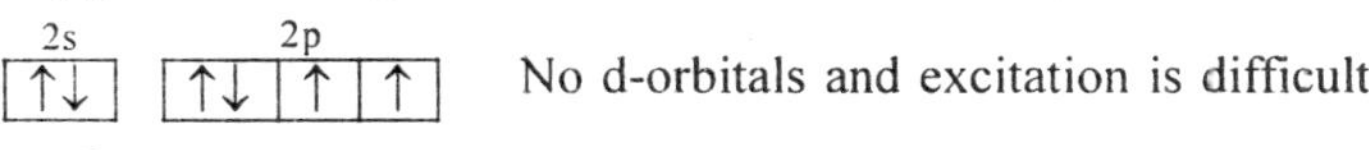

Sulphur atom in ground state

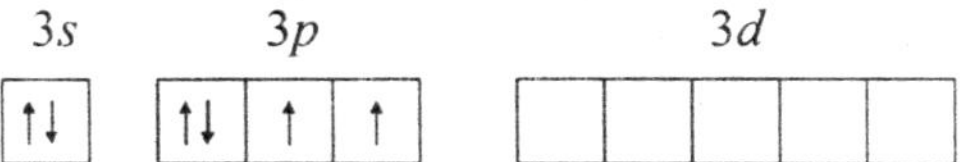

Two unpaired electrons account for a covalency of 2.

Sulphur atom in first excitation state

Four unpaired electrons account for a covalency of 4.

Sulphur atom in second excitation state

Six unpaired electrons account for a covalency of 6.

Similarly, halogens (except fluorine) also contain empty d-orbitals in their valency shell. These empty d-orbitals allow for easy excitation of p and s electrons and explain for higher valencies of 3, 5 and 7 in case of halogens.

Chlorine in ground state

One unpaired electrons account for a covalency of one.

First excitation state

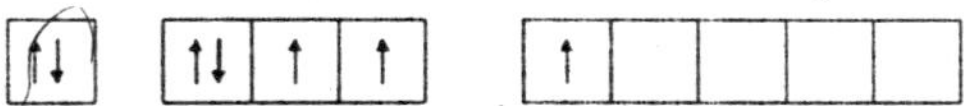

Three unpaired electrons account for a covalency of three.

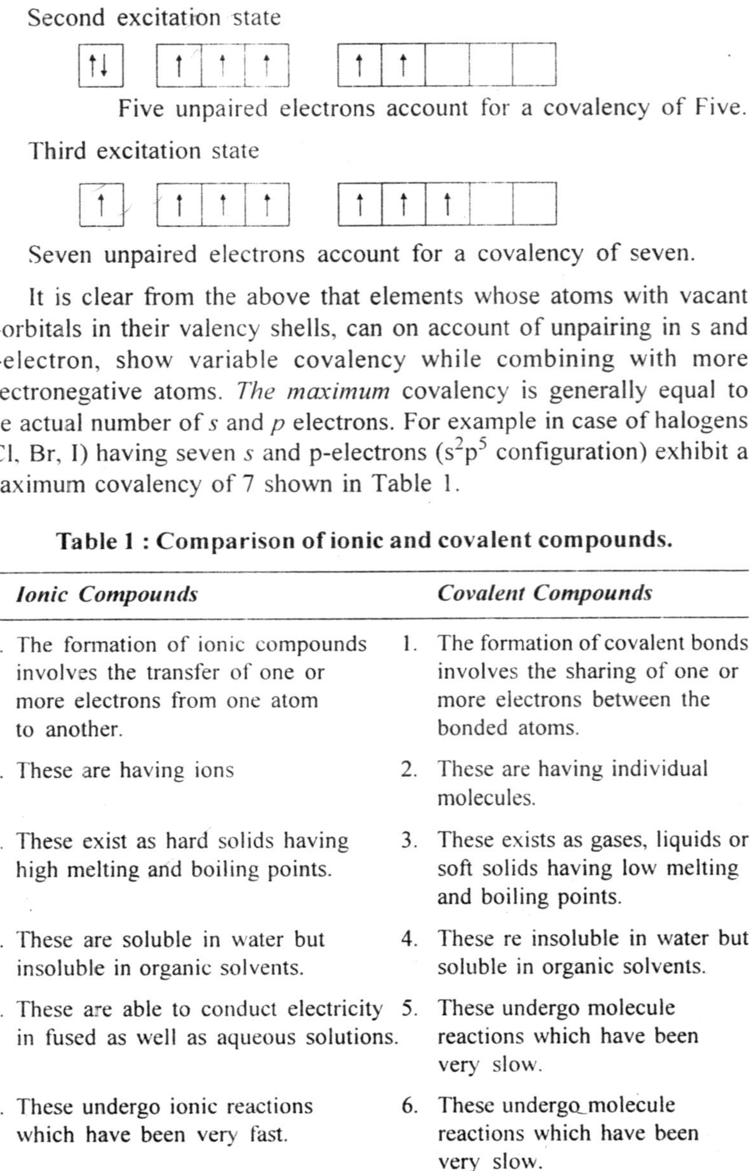

Second excitation state

Five unpaired electrons account for a covalency of Five.

Third excitation state

Seven unpaired electrons account for a covalency of seven.

It is clear from the above that elements whose atoms with vacant d-orbitals in their valency shells, can on account of unpairing in s and p-electron, show variable covalency while combining with more electronegative atoms. *The maximum* covalency is generally equal to the actual number of *s* and *p* electrons. For example in case of halogens (Cl, Br, I) having seven *s* and p-electrons (s^2p^5 configuration) exhibit a maximum covalency of 7 shown in Table 1.

Table 1 : Comparison of ionic and covalent compounds.

Ionic Compounds	*Covalent Compounds*
1. The formation of ionic compounds involves the transfer of one or more electrons from one atom to another.	1. The formation of covalent bonds involves the sharing of one or more electrons between the bonded atoms.
2. These are having ions	2. These are having individual molecules.
3. These exist as hard solids having high melting and boiling points.	3. These exists as gases, liquids or soft solids having low melting and boiling points.
4. These are soluble in water but insoluble in organic solvents.	4. These re insoluble in water but soluble in organic solvents.
5. These are able to conduct electricity in fused as well as aqueous solutions.	5. These undergo molecule reactions which have been very slow.
6. These undergo ionic reactions which have been very fast.	6. These undergo molecule reactions which have been very slow.
7. These do not exhibit isomerism.	7. These exhibit isomerism.

POLARITY OF BONDS

A covalent bond may be non-polar or partially ionic in character depending upon the electronegativities of the bonded atoms. An ionic bond on the other hand, may be thought of as 100% ionic bond. In actual practice no bond is 100% ionic or 100% covalent. But it is a common practice to term a bond as ionic if it has 50% or more ionic character. Similar a bond which has less than 5% ionic character is usually termed as covalent bond.

Covalent bonds are of two types :

(i) Non polar covalent bond.

(ii) Polar covalent bond.

(i) Non-polar covalent bond : In the non-polar covalent bonds', the two atoms have equal electronegativity, *i.e.*, equally attract the pair of shared electrons. Thus, the electron pair is shared equally between two atoms, *e.g.*, the covalent bonds in H_2, Cl_2 and F_2 are non-polar. As a result, the molecule is neutral or non-polar.

$$\underset{\text{Hydrogen molecule}}{H:H} \qquad \underset{\text{Chlorine molecule}}{:\ddot{\underset{..}{Cl}}:\ddot{\underset{..}{Cl}}:} \qquad \underset{\text{Fluorine molecule}}{:\ddot{\underset{..}{F}}:\ddot{\underset{..}{F}}:}$$

(ii) Polar covalent bond : In the case of polar covalent bonds, the two atoms have unequal electronegativities or the electron pair is shared unequally. An example is HCl, the electron pair is more attracted or displaced towards chlorine as it is more electronegative than hydrogen. Due to this, the chlorine end of the molecules appears negative and the hydrogen end positive. Such molecules having such two oppositely charged poles are called *polar molecules* and the bond is said to be *polar covalent bond.*

$$H:\ddot{\underset{..}{Cl}}: \text{ or } \overset{\delta+}{H} - \overset{\delta-}{H}$$

Similarly, BrCl molecule is a polar molecule as chlorine is more electgronegative than bromine. The shared electron pair is more attracted towards chlorine than towards bromine. As a result chlorine end of the molecule is negative and bromine end is positive.

$$:\ddot{\underset{..}{Br}}:\ddot{\underset{..}{Cl}}: \text{ or } \overset{\delta+}{Br} - \overset{\delta-}{Cl}$$

Water and ammonia afford simple examples of polar molecules containing more than one polar bond.

The ionic character of the covalent bond depends upon the electronegativity differences (Δx) of the bonded atoms. Pauling plotted Ax against per cent ionic character, P and drived the following relation.

$$P = 100\ [1 - \exp\frac{1}{4}\ (x_1 - x_2)^2]$$

However, Hannry and Symth proposed the following empirical relation :

$$P = 100(x_1 - x_2) + 3.5(x_1 - x_2)^2 \text{ (where } x_1 > x_2)$$

DIPOLE MOMENT

We have discussed earlier that two atoms differing in their electronegativity give a molecule which is positive at one ene and negative at the other end. The molecule is said to possess an *electric dipole.* The degree of polarity in a polar molecule is expressed by its dipole moment which is equal to the *product of charge and the distance between atoms.*

Mathematically – Dipole moment, $\mu = e \times d$ ...(1)

where e = electronic charge and is of the order of 10^{-10} e.s.u. and

d = distance and is of the order 10^{-18} cm.

μ is, therefore, of the order of 10^{-18} e.s.u. per cm. and this unit is known as Debye (D).

Dipole moment is a vector quantity having both direction and magnitude and is often indicated by an arrow pointing towards the negative end *e.g.*, + → –.

The molecular dipole moment of molecules having one polar bond is the same as that of the individual bonds while in case of molecules containing more than one polar bonds, the molecular dipole moment is the sum of the dipole moments of all the individual bonds. Dipole moment can be *applied* to find out the shape of the molecules, *i.e., whether symmertical or unsymmetrical.*

In general a polar bond is formed between two atoms of different radii and different electronegativities. It should, be remembered that greater the value of the dipole moment, greater will be the polarity of bond.

The following points should be kept in mind regarding dipole moments.

(i) In case of a molecule having two more polar bonds, its dipole moment is obtained by the vectorial addition of the dipole moments of the constituent bonds.

(ii) A symmetrical molecule will have dipole moment zero even though it contains polar bonds. For example, carbon dioxide, methane and carbon tetrachloride being symmetrical molecules will have zero dipole moments.

Dipole moment of methyle chloride is a vectorial addition of dipole moments of three C–H bonds and one C–Cl bond. The dipole moments of some compounds are given in Table 2.

It is found that in a polar covalent bond between two atoms, there is a partial separation of charge. This bond is, therefore, said to have a partial ionic character. The partial polar or ionic character of a bond is determined by the electronegativity difference of the atoms involved in the bond formation. Table 2 gives an approximate idea of the increase in the difference in electronegativity.

Table 2

Compounds	*Molecular dipole moment (D)*
H_2, O_2, Cl_2	0
CH_4, CCl_4, CO_2, Br_2	0
HF	1.91
HCl	1.03
HBr	0.78
HI	0.38
H_2O	1.84
H_2S	6.92
SO_2	0.60
NH_3	1.46
CK_3Cl	1.86

(a) Dipole moment of diatomic molecules : For diatomic molecules (*i.e.*, molecules have only one polar bond), the dipole moment of the molecule is the same as that of individual bonds. This is known as molecular dipole moment. For instance, the molecular dipole moment of HCl is the same as that of the single H–Cl bond, *i.e.*, 1.03 D.

(b) Dipole moment of molecules having more than two atoms: In case of molecules containing more than two atoms (or in other words having more than one polar bond), the molecular dipole moment is made up of the dipole moments of all the individual bonds.

Its magnitude depends not only on the values of individual dipole moments but also on the arrangement of the polar bonds in space. This can be understood by the following examples.

$$\overset{+\delta}{\mathrm{H}} \text{———} \overset{-\delta}{\mathrm{H}}$$

$$\mu = 1.03 \text{ D}$$

(a) Carbon dioxide molecule : As the electronegativity of oxygen is more than that of carbon, it means that carbon-oxygen bonds in carbon dioxide are polar, *i.e.*, each shared electron pair lies closer to each oxygen atom than to the central carbon atom, as shown below.

$$:\ddot{\mathrm{O}}::\mathrm{C}::\ddot{\mathrm{O}}: \text{ or } \overset{-\delta}{\mathrm{O}} = \overset{+\delta+\delta}{\mathrm{C}} = \overset{-\delta}{\mathrm{O}}$$

Experimentally, it is proved that the molecule is non-polar and its dipole moment is zero. These facts reveal that carbon dioxide molecule is linear be cause the dipole moment on one side of the molecule is cancelled by that on the other side.

(b) Carbon disulphide molecule : As the electronegativity of sulphur is more than that of carbon, it means that each shared electron pair lies closer to each sulphur atom that to the central atom and therefore both the carbon-sulphur bonds are polar.

$$:\ddot{\mathrm{S}}::\mathrm{C}:\ :\ddot{\mathrm{S}}:$$

$$\overset{-\delta}{\mathrm{S}} = \overset{+\delta+\delta}{\mathrm{C}} = \overset{-\delta}{\mathrm{S}}$$

Experimentally, it is proved that the molecule is non-polar and its dipole moment is zero. These facts reveal that carbon dioxide molecule is linear because the dipole moment on one side of the molecule is cancelled by that on the other side of the molecule.

(c) Water molecule : As each shared electron pair lies closer to oxygen atom that to hydrogen atom, it means that both the oxygen-hydrogen bonds are polar.

$$H:\ddot{\underset{..}{O}}:H \quad \text{or} \quad \overset{+\delta}{H} - \overset{-\delta-\delta}{O} - \overset{+\delta}{H}$$

Experimentally, it is proved that the molecule is polar and its dipole moment is 1.84.D. Therefore, the water molecule is non-linear. In other words, the water molecule has an unsymmetrical or bent stricture. The angle between two bonds in H_2O should be 104.5° to explain that the dipole moment of each O–H bond is 1.60 D and the net dipole moment is 1.4D. The bond angle of 104.5° has been confirmed by the X-ray analysis.

(d) Sulphur dioxide molecule : Its dipole moment is 1.60 D. It reveals that the molecule has a bent structure because the two oxygen atoms are not lying symmetrically with respect to the sulphur atom as O = S = O.

```
   S
 //  \\
O      O
```

(e) Ammonia molecule : Its dipole *moment* is 1.46 D. This value is indicative of the fact that the three hydrogen atoms are not lying symmetrically with respect to the nitrogen atom. It means that the NH_3 molecule is highly polar. The high polarity of NH_3 molecule has been explained by assigning a triangular pyramidal structure to the molecule in which nitrogen atom lies at the apex of the pyramid while the three hydrogen atoms are lying at the other three corners. Similarly, the molecules of PCl_3 and $AsCl_3$ have triangular pyramidal structures with P or As atom at the apex.

(f) BF_3, CH_4 and CCl_4 molecules : In BF_3, three are three BF bonds which are oriented at an angle of 120° to one another. The three B–F bonds lie in one plane and cancel the dipole moment of one another. Hence, the dipole moment of BF_3 would be zero. Similarly, the dipole moments of methane (CH_4) and carbon tetrachloride (CCl_4) molecules would be zero due to their symmetrical tetrahedral shape.

From the above examples it is evident that the presence of polar bonds in a polyatomic molecule does not imply that the molecule as a hole will be having dipole moment. In case of molecules such as BF_3, CH_4, CCl_4 the arrangement of individual bonds in the molecule has been such that the net dipole moment of the molecule would be zero.

Significance of Dipole Moment

1. **Knowledge of shape of molecule :** (a) Molecules possessing regular geometry are having zero dipole moment *e.g.*, carbon tetrachloride (CCl_4), methane (CH_4) etc., are having zero dipole moment. Their shape would be regular tetrahedral.
2. **Knowledge of polar and non-polar molecules:**
   (a) If a molecule is having some dipole moment, it is a polar molecule. H_2O (μ = 1.84 D) and NH_3 (μ = 1.46 D) are polar molecules.
   (b) If a molecule is having zero dipole moment, it is a non-polar molecule *e.g.*, H_2, N_2, D_2, (each has μ = 0) etc., are non-polar molecules.
3. **Knowledge of degree of polarity of molecules :** If the dipole moment of a molecule is having a high value, its degree of polarity would be high, *e.g.*, HF (μ = 1.91 D). Thus, HF is more polar than HCl. (μ = 1.03 D). Thus, HF is more polar than HCl.
4. **Knowledge of polarity in polyatomic molecules :** Polyatomic molecules have been of two types,
   (i) Molecules with symmetrical shape.
   (ii) Molecules with unsymmetrical shape.

(i) Symmetric molecules : Such molecules are said to be non-polar. In such cases, the polarity of one bond would be able too cancel the polarity of other bond, such molecules have zero dipole moment.

(i) CO_2 *i.e.* O = C = O

(ii) CCl_4 *i.e.*

(iii) CS_2 *i.e.* S = C = S

(ii) Unsymmetric molecules : Such molecules are said to be polar. In such cases, the polarity of one bond fails to cancel the polarity of other bond, *e.g.*,

(i) H_2O (ii) NH_3

Dipole Moment and Percentage Ionic Character

The concept of dipole moment has been found to be quite useful for finding the percentage ionic character in a molecule. It is demonstrated as follows.

The relation between dipole moment (μ) and percent ionic character (P) is as follows:

$$\mu = Pd/100$$

The above relation is quite useful for knowing the per cent ionic character of a bond from dipole moment (μ) and bond length (d) (Table 3).

Table 3 : Ionic Character of the H-X bonds.

Example of the molecule	*Dipole moment (Debye)*	*Bond length (pm)*	*Percent ionic character*
HI	0.38	171	5
HBr	0.78	141	17
HCl	1.03	127	12
HF	1.91	92	45

Measurement of Dipole Moment : It is possible to measure dipole moment from the relative dielectric constant, $\epsilon_r = \frac{\epsilon_r - 1}{\epsilon_r + 2} \cdot \frac{M}{\rho}$

M and ρ are molecular weight and density of the compound respectively. It is possible to obtain values of a and m from the measurements of the dielectric constants of the compound at different temperatures. Debye deduced the following relation.

$$P_m = \frac{N}{3}\left\{\alpha + \frac{\mu^2}{3\epsilon_0 kT}\right\} = \frac{\epsilon_r - 1}{\epsilon_r + 2} \cdot \frac{M}{\rho} \quad \text{(in SI units)}$$

$$P_m = \frac{4\pi N}{3}\left\{\alpha + \frac{\mu^2}{3kT}\right\} = \frac{\epsilon_r - 1}{\epsilon_r + 2} \cdot \frac{M}{\rho} \quad \text{(in C.G.S.) units}$$

Applications of Dipole Moment : The various applications of dipole moments are as follows.

1. **Bond polarity :** By using equations as given above it is possible to determine the electronegativity difference of the bonded atoms from the dipole moment measurements. When dipole moment is referred to a bond in a molecule, it is known as the bond moment. Some typical values for bond moments in Debye are given in Table 4.

Table 4 : Bond moments in Debyes.

Bond	*Bond moment (Debye)*	*Bond*	*Bond moment (Debye)*
H–C	0.40	C–F	1.39
H–N	1.33	C–Cl	1.47
N–O	1.36	N–F	0.17
H–P	0.36	C=N	0.90
H–S	0.68	C=O	2.3
C–N	0.23	C°N	3.5
C–O	0.74	N=C	3.0
C–S	0.90	$N^+ - B^-$	2.55
		$N^+ - O^-$	4.3

2. **Cis and trans isomers :** Dipole moment has been used to characterise cis and trans isomers. For example, the cis isomer of a square planar complex has dipole moment whereas the trans isomer does not have.

3. **Structure equilibria of complexes :** Measurement of dipole moments have been used for the determination of concentration of tetrahedral and square planar complexes at equilibria. These two forms have different dipole moments, due to different stereochemistries.

LIMITATIONS OF THE ELECTRONIC THEORY OF VALENCY

It will appear from the above that the octet theory provided a general explanation in terms of electrons, for the three types of valency bonds. This theory, however, could not cope with the growing knowledge on the size and shape of molecules and the nature and strength of valency bonds. Some of these are illustrated as follows :

(a) Molecules in which the central atom has less than eight electrons : There are a large number of stable compounds which possess an insufficient number of electrons than allowed by octet theory.

An interesting example is B_2H_6 in which two boron atoms each possessing three electrons can share these with 6 hydrogen atoms without leaving any electron for the bond between the two boron atoms.

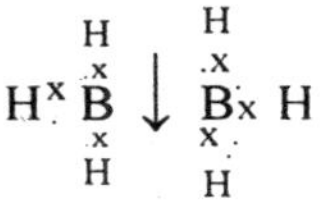

H H
.x .x
H^x B ↓ Bx H
.x x .
H H

Another example is $BeCl_2$ in which the central Be atom has only four electrons in its valency shell whereas each of the Cl atoms is obeying the octet rule. Be $(C_2 H_5)_2$ is also an example of this type. One more example is BF_3 in which the central atom has only six electrons in its valency shell. Other similar examples are BCl_3, B $(CH_3O)_3$ etc.

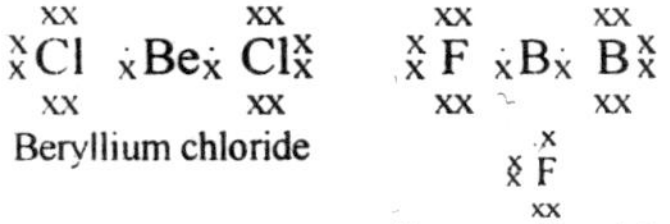

xx xx xx xx
^x_xCl xBex Cl^x_x ^x_xF xBx B^x_x
xx xx xx xx
Beryllium chloride

x
^x_xF
xx
Boron trifluoride

(b) Molecules in which the central atom has more than eight electrons : In the octet theory it is assumed that the maximum number of electrons in the valency shell is 8. However a large number of molecules are known in which the central atom has more than eight electrons. For example, in PCl_5 the phosphorus atom has ten electrons in its valency shell. Similarly, Cl in DCl_3 and I in ICl_3 have ten electrons in their valence shell.

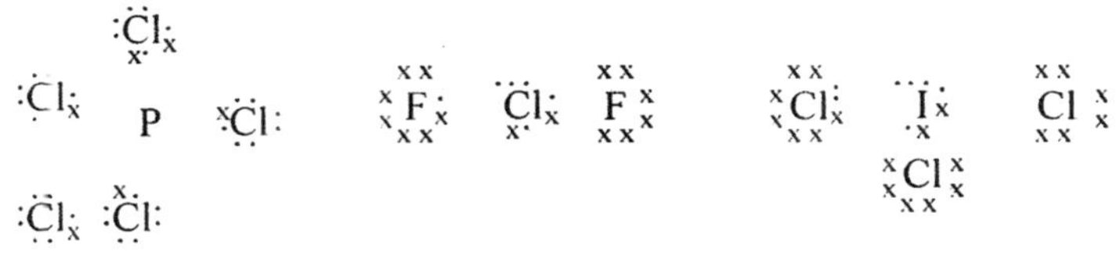

:Cl:x

:Cl:x P xCl: ^{x}F: Cl:x F^x_x ^{x}Cl: I:x Cl ^x_x

:Cl:x :Cl: xCl^x_x

Fig. 1(a)

In SF_6, sulphur is surrounded by twelve electrons. In OsO_4 and OsF_8, Os has sixteen electrons in its valence shell.

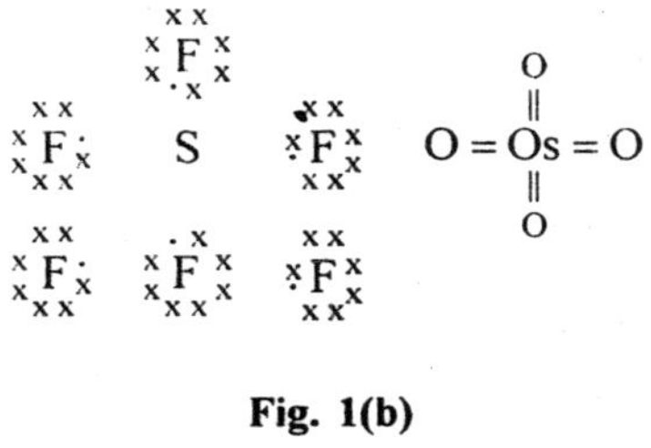

xx
^x_xF^x_x
xx xx xx
^x_xF: S xF^x_x O = Os = O
xx xx
xx . x xx
^x_xF^x ^x_xF^x_x xF^x_x

O
‖
O = Os = O
‖
O

Fig. 1(b)

(c) The octet theory did not offer any explanation for the formation of one-electron bonds as in H_2^+ as well as 3-electron bonds as are present in NO, O_3, NO_2.

(d) The octet theory cannot explain the formation of metallic bonds.

(e) Transition metal ions often show variable valency by flouting the octet rule.

COVALENT BOND NATURE

This is important to remark that the treatment of the covalent bond is far more difficult than the electrovalent bond. In the covalent bond one has to understand what types of forces exist which keep the electron pairs in contact with the two atoms. This idea has been explained on the basis of *Wave Mechanics*. Two wave mechanical theories of the covalent bond have been formulated to explain the nature of the covalent bond. These approaches are.

(A) Valence Bond Method : It is due to the work of *Heitler, London, Slater and Pauling*. In this treatment, a molecule is formed by the overlapping of atomic orbitals and a *molecule is composed of atoms which retain their individual character when linked to the other atoms.*

(B) Molecular Orbital Method : It is due to the work of *Hund* and *Mulliken*. In this treatment, atoms composing of molecular orbitals lose their individual character, *i.e., an electron moves in the field of more than one nucleus. It means that molecular orbitals are polycentric.*

SIGMA (σ) AND PI (π) BONDS

A covalent bond or molecular orbital is of two types.

(i) Sigma (σ) bond

(ii) Pi (π) bond

When a bond is formed between two atoms by the overlap of orbitals along their axis (end–to–end–overlap), the bond formed is called *sigma* (σ) bond.

At the electron clouds overlap along their axis, maximum overlap is possible. Hence, the sigma (σ) bond is a strong bond. Further, electron cloud of a σ bond is symmetrical about the line joining two nuclei concerned. Hence s bond is defined as :

"Bond orbitals which are symmetrical about the line joining the two nuclei are known as s bonds".

These types of bonds are explained below :

(1) Sigma (s) bond : *This type of bond gets formed by the heat on overlapping of half-filled atomic orbitals along the inter-nuclear axes. It is a* **strong bond** *as the extent of overlapping is maximum.*

Types of axial overlap : Various types of axial overlap of s and p electronic orbitals are as follows :

(i) s–s orbitals overlapping

(ii) s–p orbitals overlapping

(iii) s–s orbitals overlapping.

(i) s–s orbitals overlapping or s–s sigma bond : This type of sigma bond is obtained when half-filled s–orbitals of two atoms get overlapped. For example, **the formation of hydrogen molecule** by the overlapping of 1s–orbitals of two hydrogen atoms, has been involving this type of bonding.

(ii) s–p–orbitals overlapping or s–p sigma bond : This type of sigma bond is obtained when half filled s–orbital of one atom gets overlapped with half–filled orbital of other atom. Following examples are given to illustrate this concept :

1. Formation of Hydrogen fluoride (HF)

$_9F = 1s^2\ 2s^2\ 2p_x^{\ 2}\ 2p_y^{\ 1}\ 2p_z^{\ 1}$. It is having one half filled $2p_z$ orbital.

$_1H = 1s^1$. It is having one half filled s–orbital.

One lobe of half filled 2pz orbital (dumbell in shape) gets overlapped with s–orbital of H–atom along the internuclear axis. Because of this sp–molecular orbital is produced. Due to higher electronegativity of F–atom than that of H–atom, the bond pairs of electrons reside more closer to F–atom. Because of this, charge separation occurs.

2. Formation of water (H_2O molecule) : O atom ($1s^2\ 2s^2\ 2p_x^{\ 2}\ 2p_y^{\ 1}\ 2p_{z1}$) is having two half filled orbitals while H–atom (1s1) is having one half filled s–orbital. Half filled 2p orbitals of oxygen undergo overlapping with s-orbitals of each H–atom to form sp sigma bonds.

The p-orbitals are making an angle of 90° with one another yet the bond angle in water is 105°. It occurs due to the repulsive interaction between lone pairs and also between lone pairs and bond pairs.

3. Formation of ammonia (NH_3) molecule : N atom ($1s^2\ 2s^2\ 2p_x^1\ 2p_y^1\ 2p_{z1}$) is having three half filled 2p orbitals. 1H atom ($1s^1$) has one half filled s–orbital. Half filled 2p-orbitals of nitrogen get overlapped with s–orbital of each H–atom to yield s–p sigma bonds. The p–orbitals are making an angle of 90° with one another. But the bond angle in NH_3 is 108°. It is attributed to the repulsive interactions between lone pair–bond pair–bond pair.

(iii) p–p orbitals overlapping or pp–p sigma bond : This type of sigma bond is obtained when half filled p orbital of one atom gets overlapped with half filled p-orbital of other atom. Following example is given for illustration :

Formation of Fluorine Molecule

($_9F$ = 1s2 $2s^2\ 2p_x^2\ 2p_y^2\ 2p_z^1$). One lobe of half filled 2p–orbital of one F–atom is overlapping with one lobe of half filled $2p_z$–orbital of second F–atom along the internuclear axis. Because of this, p-p (σ) molecular orbital is obtained.

2. Pi (π) bond : *This type of bond is obtained by the sidewise overlapping of half filled atomic orbitals in a direction which is perpendicular to the internuclear axes.* It is a *weak bond* as the extent of overlapping of atomic orbitals has been somewhat small.

Type of sideways or lateral overlap : The only type of side–wave overlapping has been the p–p atomic orbital overlapping which is perpendicular to the inter nuclear axes. Following reveals that the two p-orbitals have been perpendicular to the inter-nuclear axes. On coming closer, these slightly overlap laterally. The upper lobes of the two orbitals tend to overlap above and lower lobes below the inter-nuclear axes. Pi-molecular orbital is having electron cloud above and below the inter-nuclear axis. Following examples are given for illustration.

(1) Formation of oxygen (O2) molecule : O–atom ($1s^2\ 2s^2\ 2p_x^2\ 2p_y^1\ 2p_z^1$) is having two half filled p–orbitals. One half-filled 2p-orbital of one O-atom is overlapping with one half-filled 2p-orbital of another O-atom along inter-nuclear axis to form σ-bond.

Second half-filled 2p orbital of one O-atom is overlapping sideways with second half-filled 2p-orbital of another O-atom to yield p-bond. The molecule is a double bond which consists of one sigma and one p-bond. The covalency of oxygen would be two.

(3) Formation of nitrogen (N_2) molecule : N-atom ($1s^2 2s^2 2p_x^1 2p_y^2 2p_z^1$) is having three filled 2p-orbitals. One of these half-filled p-orbitals of one N-atom would overlap with one half-filled 2p orbital of another N-atom along inter-nuclear axis to yield σ bond. Other two half-filled 2p orbitals of one N-atom would be overlapping sideways with each half-filled 2p-orbital of another N–atom to yield two p-bonds. The molecule has a triple bond consisting of one sigma and two o-bonds. The covalency of nitrogen would be three.

Characteristics of sigma (σ) and Pi (π) bonds : The characteristics of sigma and pi bond are given below under therein Table 5.

Table 5 : Comparison of sigma (σ) and PI (π) bond.

Sigma (σ) bond	*Pi (π) bond*
(i) It is a strong bond as orbitals overlap to greater extent.	(i) It is a weak bond as orbitals overlap to less extent.
(ii) s-orbitals are involved in the formation of a sigma bond.	(ii) s-orbitals are not involved in formation of pi bond.
(iii) One lobe of p-orbital involving in s-bond formation stretches whereas the other lobe reduces in size.	(iii) Both lobes of p-orbitals involving in a p-bond formation remain of the same size.
(iv) This bond is formed by overlap of orbitals along their inter-nuclear axis, *i.e.*, end to end overlapping.	(iv) Pi bond is formed by side to side overlapping of orbitals (Lateral overlapping)
(v) Sigma bond is a strong bond.	(v) Pi bond is a weak bond.
(vi) Electron cloud of a sigma bond is symmetrical about the line joining two nuclei concerned.	(vi) Electron cloud of pi bond is asymmetrical.
(vii) There can be free rotation of atoms around the sigma bond.	(vii) As the electron clouds overlap above and below the plane of the atoms free rotation is not possible around a pi bond.
(viii) It determines the direction and extent of internuclear distance.	(viii) It has no primary effect on the direction of bond but shortens the internuclear distance.

GEOMETRY AND SHAPES OF MOLECULES

Introduction

Molecules known to occupy three dimensions are having particulars shapes which are determined by the bonding arrangements of the atoms. The shape of a molecule becomes important in ascertaining its chemical and physical properties, *e.g.*, methane (CH_4) and water (H_2O) are having nearly equal molecular weights but- methane boils at –161° C and water boils at 100° C. Much of this great difference in boiling points has been attributed to the difference in the shapes of these molecules. By the geometrical structure of a molecule, it is meant the three dimensional arrangement of the atoms of groups.

Ionic Compounds are made up of cations and anions which are held together by electrostatic forces of attraction to form a crystal or giant molecule. In a crystal, each cation gets surrounded by a definite number of anions and vice versa. This number has been found to depend upon the size and charge on the ion and decides the geometry of the molecule. An example is that, sodium chloride does not exist as Na^+ Cl^- but in this crystal, each Na^+ ion gets surrounded by six Cl^- – ions and *vice versa*.

Because of this, sodium chloride is having octahedral geometry. In other words, *ionic compounds in the solid state do not exist as simple molecules but the ions are arranged themselves in a definite three dimensional pattern called lattice.*

On the other hand *Covalent Compounds* generally are made up of simple individual molecules like O_2, CO_2 and HCl. However, in these simple covalent molecules the constituent atoms have been occupying definite positions relative to one another.

The shape of a molecule depending on the arrangement of valencies within the molecule is called the geometry of the molecule. For example, a diatomic molecule, formed by the bonding of two atoms, has a linear shape. In case of other simple molecules, there is always a central atom which forms chemical bonds with other atoms. There are various possibilities in which these bonds may have their mutual *orientations* (or directions) with respect to one another and with respect to central atom.

Thus, depending upon their relative orientations, the molecule, as a whole will have different shapes. Since their shapes may be different they will possess different bond angles between a pair of bonds and bond lengths. When the bonded atoms are joined by imaginary lines, a space model is obtained. This is termed as the shape of the molecule. Each molecule is having a definite geometry.

Reasons for Definite Geometry In Molecules

Atoms react to form molecules so as to attain a state of minimum energy. An atom gets bonded to other atom in a molecule by a pair of electrons between the nuclei of bonded atoms. In polyatomic molecule, there exists more than one electron pair around the central atom. These electron pairs tend to repel each other, as they are similarly charged. As the repulsive interactions gives rise to an increase in energy, the electron pairs will try to remain as far away from one another as possible. Because of this.

(i) All bonded atoms have been occupying such positions (around the central atom) in space where the repulsive forces between them have been minimum.

(ii) The molecule tends to attain minimum energy and maximum stability.

As there can be only one definite orientation of bonded atoms (or electron pairs) around the central atom corresponding to a state of minimum energy, a molecule would be having only one fixed shape or definite geometry.

In order to decide the geometry of the various molecules, various treatments are employed which are as follows :

1. Helferich's Rules
2. VSEPR Theory
3. Hybridisation.

We will now discuss these one by one.

1. Herferich's Rules : A useful set of empirical rules, called *Helferich's rules,* can be used to arrive at the spatial arrangement of atoms in simple molecules. *In other words, these rules can be used to determine or predict the shapes of simple molecules.* These rules may be summarised as follows :

(*n*) *Molecules or ions of type AX_2 where X may be the same as A or different.* If the central atom A has no unshared electrons, the molecules or ions of the type AX_2 have straight-line structure. For example, carbon dioxide (CO_2) nitrous oxide (N_2O) and oxide (N_3^-) ion have liner structures.

$$:\ddot{O}: :C::\ddot{O}: :\ddot{N}::\overset{+}{N}:\overset{-}{\ddot{O}}: :\overset{-}{\ddot{N}}::\overset{+}{N}: :\overset{-}{\ddot{N}}:$$

If the central atom A has one or more unshared electron pairs, the molecules or ions of the type AX_2 will have angular shapes.

For example, water (H_2O), hydrogen sulphide (H_2S), and ptannous choride ($SnCl_2$) should be angular, as the following formulae show :

(b) *Molecules or ions of the type AX_3 where X is different or the same as A.* When a does not carry an unshared pair of H electrons, the molecules or ions AX have planar structures with A at the apex. For example, in the BH_3 or BF_3 or CO_3^{-2} OR NO_3^- species, the bond angle is 120°. But since, there is no unbonded electron pair, the angle is not decreased but remains 120°.

When A has one unshared pair or electrons, the molecules or ions of the type AX_3 have pyramidal shapes with A at the apex. For example, NH_3, SO_3^{2-} and ClO_3^- are pyramidal with bond angle of 109° 28' but the unbonded pairs tend to occupy more space, resulting in decrease in the bond angle between bonded pairs. In NH_3, the N–H bonds lie at 107° to each other.

Molecules or ions of the type AX_4; Molecules or ions of the type AX_4, have tetrahedral structure if the central atom A has no unshared electrons in its other shell and if this shell is of the rare-gas type. If the outer shell is not of the rare-gas type, molecules or ions of the type AX_4 have planar structure.

For example, methane (CH_4) and ammonium ion (NH_4^+) have tetrahedral structures whereas the four co-ordinate complexes of Ni (II) and Pd (II) have planar structure. It should be born in mind that Halferich's rules are merely empirical generalisations that are sometimes useful in arriving at the directed nature of covalent bonds in simple molecules.

2. Valence Shell Electron-Pair Repulsion Theory (VSEPR Theory): This theory was developed by *Gillespie* and *Nyholm.* According to this theory, the geometry of a molecule depends upon the number of bonding and non-bonding electron-pairs (lone pairs) in the central atom which arrange themselves in such a way that there is a minimum repulsion between them so that the molecule has minimum energy and maximum stability.

Since there can only be one orientation of orbitals corresponding to minimum energy a molecule of a given substance has invariably a definite shape *i.e.* a definite geometry. Gillespie has proposed the following rules for explaining the shapes of molecules and ions.

Rule 1 : When the central atom in a molecule is surrounded only by bonded electron pairs but not by lone pairs, the molecule will have a regular geometry of shape. The geometry depends on the number of bonded electron pairs as given in the Table 6.

Table 6

Number of electron pairs	*Geometry*	*Bond angle*	*Examples*
2	Linear	180°	BeF_2,$HgCl_2$,CdI_2
3	Trigonal plane	120°	BF_3, BI_3, GaI3
4	Tetrahedral	109° 28¢	CH_4, $SnCl_4$, $SnBr_4$, BF_4^{1-}
5	Trigonal bipyramid	90° and 110°	PCl_5, $SbCl_5$, VO_3^{-1},
6	Octahedral	90°	SF_6, $[SiF_6]^{2-}$ $[PbCl_6]^-$
7	Pentagonal bipyramid	72° and 90°	IF_7

Rule 2 : Some molecules may have distorted or irregular geometry if the electron pairs surrounding the central atom are not equivalent, *i.e.* some are bonded while remaining are lone pairs. Lone pair electrons repel adjacent electron pairs more strongly than do bonding electron pairs. Also, the repulsion is expected to increase in the following order :

bond pair — bond pair < bond pair–lone pair < lone pair-lone pair.

Rule 3 : The greater the number of lone pairs of electrons on the central atom, the greater, in general will be the tendency of bond angles to contract and hence, greater will be the distortion in the geometry of the molecule. This rule is understood by considering the following series.

$$NO_2^+, 110°; \dot{N}O_2\ 104°;\ NO_2^-, 115°$$

Rule 4 : The smaller the electronegativity of the central atom, the smaller is the bond angle and hence, smaller is the distortion in the regular shape of molecules. For example,

(i) PI_3, 102° > PBr_3, 101.5° > PCl_3, 100°

(ii) AsI_3, 101.4 > $AsBr_3$, 100.5°>$AsCl_3$, 98.4°

Rule 5 : Bond angles involving multiple bonds are generally larger than those involving only single bonds. However, multiple bonds do not affect the geometry of a molecule.

Rule 6 : Repulsion between electron pairs in filled shell is greater than that in incomplete shell. For example, the bond angle decreases.

$OH_2 \geq SH_2 > SeH_2 > TeH_2$

104.5 92.2 91.0 89.5

Reason : Lone pair of electrons may be defined as that pair which is found in the valence shell of the central atom. It is not involving in bonding (except when required) and gets influenced by the nucleus of the central atom only.

Due to the attractive influence of two nuclei the electron cloud of the shared pair "contracts" and would be occupying a smaller volume. On the other bond the electron cloud of the lone pair is under the attractive influence of only one nucleus "spreads out" and would be occupying a larger volume.

Hence, lone pairs due to their larger electron clouds become more nearer to other lone pairs. This gives rise to maximum repulsion between lone pairs. Due to the small size of the electron cloud of bond pairs, the repulsive force between the two bond pairs would be the least. The repulsion between a lone pair and bond pair would be somewhat in between the two extreme cases.

Applications of VSEPR Theory : By applying the above mentioned rules of VSEPR theory, one can deduce the regular geometry as well as the distortion in its regular shape in different molecules.

1. Regular Geometry : According to this theory, a molecule will have a regular geometry if the central atom is surrounded only by orbitals containing shared pairs of electrons (*i.e.*, the bond pairs) and there are no orbitals containing lone pairs (non-bonding) in the valency shell.

Gillespie and Nyholm calculated theoretically (with the help of simple geometry) the regular geometries of molecules containing two to eight pairs of bonding electrons in the valency shell of a central atom.

(a) Two bonding electron-pairs : If there are two bonding pairs of electrons in a molecule, the only way to keep them to the maximum extent is to arrange them at 180° to each other. It yields a linear molecule with a bond angle of 180°. Example is the molecule of beryllium fluoride.

F – Be – F.

(b) Three bonding electron-pairs : If there are three orbitals around the central atom in a molecule, the molecule in this case will be planar triangle and in this case the bond angle should be 120°. This is the case with boron trifluoride.

Here B, the central atom has the electronic configuration : $1s^2$ $2s^2$, $2p^1$ or 2, 3, *i.e.,* it has three electrons in its valency shell. These three valency electrons form three covalent bonds with three fluorine atoms.

```
  :F̤̈:         F
               |
            F – B – F
:F̤̈: B :F̤̈:
```

The molecule has the shape of an equilateral triangle with boron atom lying at the centre and the three atoms lying at the corners of the triangle. This shape is also known as trigonal plane. The bond angle F–B–F is 120°

(c) Four bonding electron-pairs : If there are four orbitals around the central atom in a molecule, the only way to keep them farthest apart is to have them directed towards the four corners of a regular tetrahedral. Thus, the molecule will be tetrahedral in shape and the bond angle in this case will be 109.5°.

An example for this is methane. Hence the central atom, C, has the electronic configuration : $1s^2$, $2s^2$, $2p^2$, outer orbit has four electrons in its valency shell . These electrons form four equivalent covalent bonds with 4 hydrogen atoms,

```
   H                H
                    |
H:C̤̈:H   or   H – C – H
                    |
   H                H
```

As all the four bond pairs are equivalent, the molecule has a regular geometry of *tetrahedron* with a bond angle H–C–H as 109° 28.

(d) Five bonding electron pairs : If there are six bonding electron pairs the only way to keep them farthest apart is to have an arrangement of the type, triogonal bipyramidal. Examples are PCl_5 (g) and PF_5.

(e) Six-bonding electrons–pairs : It there are six bonding electron pairs in a molecule, the molecule will be octahedral in shape and the bond angle will be 90°. An example of this is SF_6. The central atom, 6 flourine atoms. Since all the six bond pairs are equivalent and no valency electron is left unused, the molecule has a regular geometry of an *octahedron.*

(f) Seven bonding electron pairs. If there are seven boding electron pairs, the molecule will be pentagonal bipyramid. An example of this is IF_7. The structure of this molecule can be better understood if we visualise two pyramids placed in such a way that their bases touch one another congruently. The base of the pyramid is a pentagon.

In terms of VSEPR Theory, a pair of electrons in the valence shell, either bonding or lone pair, have been under the influence the structure of the molecule. These electron pairs, are termed as *subtractive* as they are effective in ascertaining the stereo or structural arrangement of atoms in a molecule.

However, the lone pair of electrons due to their more diffuse (spread out) electron clouds bring about larger repulsions. Because of this, bond pairs move away from them and come closer to each other. This causes the decrease in bond angle from the normal expected values. The effect of the lone pairs on molecular geometry in case of the hydrides of the second row non-metals (having lower subtractive electron pairs).

Regular Geometry of Covalent Molecules

A molecule would be having a regular geometry when the repulsive interactions between the electron pairs have been of equal magnitude.

Conditions for regular geometry : A molecule would be having a regular geometry when it satisfies the following conditions.

(i) The central atom of the molecule should be bonded to all similar atoms, *e.g.*, CH_4, CCl_4.

(ii) The central atom should be surrounded by bonded pair of electrons and should not be having lone pair of electrons.

(iii) All the bond lengths between the central atom and the bonded atoms should happen to be the same.

(iv) The central atom should get bonded to other atoms by single covalent bonds (or same type of bonds).

Distorted or Irregular Geometry of Molecules

A molecule would *be* having irregular geometry if the repulsive interactions between the electron pairs (around the central atom) have been not of equal magnitude.

Conditions for irregular geometry : A molecule is said to be having irregular geometry when it satisfies any one of the following conditions :

(i) The central atom of the molecule should not get bonded to all the similar atoms, *e.g.*, $CHCl_3$, CH_3Cl.

(ii) The central atom should possess one or more lone pair of electrons in addition to one or more bond pair of electrons, *e.g.*,

$$\ddot{N}H_3,\ H_2\ddot{\underset{..}{O}}$$

(iii) The central atom may get bonded to similar atoms but should possess different bond lengths, *e.g.*, in PF_5 molecule, the equatorial P–F bond length has been 2.04Å whereas the axial P–F bond length has been 2.19Å.

(iv) The central atom should get bonded to other atoms by single as well as multiple bonds, *e.g.*, SO_3.

Prediction of Shapes of Molecules

In order to have an idea about the geometry on the basis of VSEPR Theory, one should proceed as follows :

(1) First of all count the number of valence electrons. Then write down the Lewis structure for the molecule or ion.

(2) From the Lewis structure, find the number of unshared electron pairs and number of bonds which are made by the central atom to the other atoms (The multiple bonds are counted as a single bond).

(3) Find out the number of subtractive electrons pairs by adding the total number of shared pairs and lone pairs around the central atom. Then, predict the geometry of the molecule.

(a) Shapes of Molecules Having Regular Geometry

1. Shape of BeF_2 molecule (AB_2 type)—Linear : In BeF_2, the central atom is $_4Be$ which is having the configuration $1s^2\ 2s^1\ 2p^1$ in the excited state. It is able to form two bonds with two fluorine atoms because of the presence of two unpaired electrons in its valence shell. In the formation of BeF_2 each of these electrons gets shared by two fluorine atoms. Because of this, the beryllium atom gets surrounded by two bonded pair of electrons.

Total number of electrons present in the valence shell of Be = 2

Bonds formed by Be = 2

Number of shared pairs of electrons around Be = 2

unshared pair of electrons = 1/2 [2 – 2] = 0

Thus, total number of subtractive electron pairs around central Be atom = 2 + 0 = 2

But the central Be atom gets bonded to similar F–atoms by single covalent bonds and get surrounded by only two shared pair of electrons. Therefore, the force of repulsion among bonded electron pairs would be equal. Hence, the molecule of BeF_2 is having regular linear geometry with bond angle as 180°.

Other examples of molecules having linear geometry include $ZnCl_2$, $MgCl_2$ $BeCl_2$, BeH_2 and $HgCl_2$.

2. Shape of BF_3 molecule (AB_2 type)—Triangular : In BF_3, central atom is $_5B$ which is having the configuration as $1s^2\ 2s^1\ 2p_x^1\ 2p_y^1$ in the excited state. It is able to make three bonds with three fluorine atoms because of the presence of three unpaired electrons in its valence shell. In the formation of BF_3, each of these electrons gets shared by three fluorine atoms. Because of this, the boron atom gets surrounded by three bond pairs of electrons.

$$\overset{\times}{B}{}^{\times}_{\times} + 3\cdot\ddot{\underset{\cdot\cdot}{F}} \longrightarrow :\ddot{\underset{\cdot\cdot}{F}}\ {}^{\times}\overset{\times}{B}{}^{\times}\ \ddot{\underset{\cdot\cdot}{F}}:\quad (\text{third } :\ddot{F}: \text{ above B})$$

But the central B atom in BF3 has been bonded to three similar fluorine atoms by single covalent bonds and has been surrounded by only three shared pairs of electrons. Therefore, the force of repulsion among bonded electron pairs would be equal. Thus, the molecule of BF3 is having regular triangular planar geometry. The three B–F bonds get directed towards the corners of an equilateral triangle with each bond angle as 120°.

Other examples of molecules having triangular planar geometry include BCl_3, BH_3 and $AlCl_3$.

3. Shape of CH_4 molecule (AB_4 type)—Tetrahedral : In CH_4 central atom is C which is having the configuration $1s^2\ 2s^1\ 2p_y^1\ 2p_z^1$ in the excited state. It is able to make four bonds with four hydrogen atoms because of the presence of four unpaired electrons in its valence shell. In the formation of CH_4 each of these electrons gets shared by four hydrogen atoms. Because of this, the carbon atom gets surrounded by four bonded pair of electrons.

$$\underset{x}{\overset{x}{xCx}} + 4\dot{H} \longrightarrow \underset{\underset{H}{.x}}{\overset{\overset{H}{.x}}{H_x\ C_x\ H}}$$

Total number of electrons present in the valence shell of carbon atom = 4.

Number of bonds made by carbon = 4

Number of shared pair of electrons around carbon = 4

Number of unshared pairs of electrons around carbon

$$= 1/2[4 - 4] = \text{Zero.}$$

Total number of stereoactive electron pairs present around carbon

$$= 4 + 0 = 4.$$

The central carbon atom in CH_4 molecule gets bonded to four similar hydrogen atoms by single covalent bonds and gets surrounded by only four shared pair of electrons. Therefore, the force of repulsion among bonded electron pair would be equal. Hence, the molecule of CH_4 is having regular tetrahedral geometry.

Four C–H bonds get directed towards the four comers of a regular tetrahedron while carbon atom is situated at the c enter of this tetrahedral, each bond angle would be 109° 28′. In CH_4 molecule three C–H bonds are situated in one plane while four the C–H bond is situated behind the plane. Other examples of molecules possessing tetrahedral geometry included CCl_4, $SiCl_4$ and $SnCl_4$.

4. Shape of SH_6 molecule (AB_6 type)—Octahedral : In SF^6, central atom is $_{16}S$ which is having the configuration $1s^2\ 2s^2\ 2p^6\ 3s^1\ 3p_x^1\ 3p_y^1$

3pz1 $5d^2$ in the excited state. It is able to form six bonds with six fluorine atoms because the presence of six unpaired electrons in its valence shell. In the formation of SF_6, each of these electrons gets shared by six fluorine atoms. Because of this, the sulphur atom gets surrounded by six bounded pair of electrons.

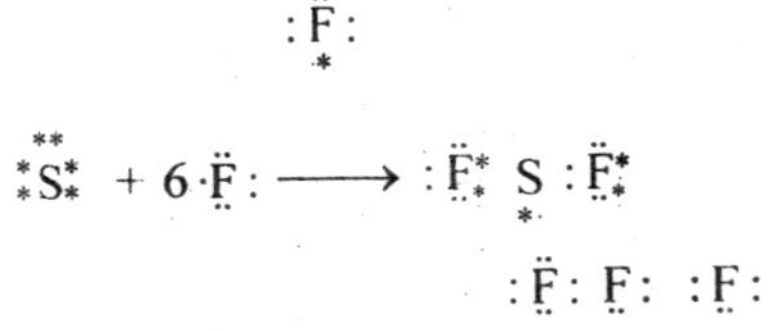

:F̈:

S + 6·F̈: ⟶ :F̈* S :F̈*

:F̈: F: :F:

Total number of electrons present in the valence shell of S = 6

Number of bonds made by sulphur = 6

Number of shared pairs of electrons around S = 6

Number of unshared pairs of electrons around S = 1/2 [6 – 6] = 0

Total number of stereoactive pairs present around S = 6 + 0 = 6

But the central sulphur atom in SF_6 gets bonded to six similar fluorine atoms ad gets surrounded by only six shared pairs of electrons. Therefore the force of repulsion among bonded electron pairs would be equal.

Thus, the molecule of SF_6 is having regular octahedral geometry. In such a geometry, the central sulphur atom is found in the centre of a square. Out of the remaining two S–F bonds, one is situated above and the other is situated below the square plane and are also perpendicular to it, F–S–F bond angle has been 90°.

Other example of a molecule having octahedral geometry has been TeF_6.

Bond angle in H_2O is less than that in NH_3 : In water molecule two lone pairs of electrons on oxygen are present. In NH_3 molecule, one lone pair of electrons on nitrogen is present. According to VSEPR theory, lone pair-lone pair repulsion > lone pair bond pair repulsion > bond pair-bond pair repulsion. As the number of lone pairs have been more in oxygen, the O–H bond pairs come more closer than N–H bond pairs in NH_3. Hence, bond angle in H_2O (104.5°) would be less than in NH_3 (107°).

Bond angles in the hydrides of elements belonging to the same group : In case of compounds which are differing only in the nature of central atom (the number of electron pairs surrounding central atom and the nature of attached atoms being the same), the bond angles get decreased on moving down the group. For example, the bond angles in the case of hydrides of sixth group elements have been as follows :

Hydride	H_2O	H_2S	H_2Se	H_2Te
Bond angle	104.5°	92.6°	91.0°	90.0°.

The central atom in each case gets surrounded by two bonded pairs and two lone pairs of electrons. Thus, each molecule would be V shaped but bond angles would be different. The reason is that when we move down the group, the size of the central atom goes on increasing and its electronegativity goes on decreasing.

Because of this, the bond pairs of electrons lie farther away from the central atom, therefore the force of repulsion between the bond pairs gets decreased. The bond pairs gets decreased. The orbitals having bond pairs will attempt to come closer. Hence, bond angles show the corresponding decrease. Similarly, it is also possible to explain the decrease in the bond angles in the hydrides of V group elements. The bond angles in these are given below :

Hydride	NH_3	PH_3	AsH_3	SbH_4
Bond angle	107°	93.3°	91.8°	91.3°

2. Shape of BrF_3 molecule : In this molecule, there are ten valency electrons in the valency shell of the bromine atom. Out of these, six electrons form three bonding pairs whereas the remaining four electrons form two lone pairs. The expected geometry of BrF_3 is trigonal. But the presence of two lone pairs yields a trigonal bipyramid structure with the lone pairs in equational position. The mutual repulsion of the lone pair binds the axial fluorine atom to give a 'bent T" structure.

3. Shape of ICI_4^- : In this molecule, there are twelve electrons in the valency shell of iodine. Out of these, eight electrons form four bonding electron pairs whereas the remaining four electrons form two lone pairs.

This molecule should have an octahedral geometry with two positions occupied by the lone pairs which can be in the *cis* position or in *trans* positions does not give as great angular volume as the *trans* position does. However, the structure of ICI_4^- is square planar with the lone pairs above and below the plane, which are however not shown in the diagram.

Presence of Three Lone Pairs

Formation of ICl_2^- **:** On the face of it the ICl_2^- anion should have a linear structure. Actually this is found to be so. Although there are three lone pairs, they arrange them-selves at the equatorial positions, causing no distortion. In the centre of this diagram there is iodine atom.

Stability and less reactivity of SF_6 **:** In SF_6 molecule, all the S–F bonds have been equivalent. The repulsive force between bonded pairs force all atoms to occupy such positions (around the central atom) in space where the repulsive forces between them become equal and minimum. The molecule, therefore, gets minimum energy and maximum stability.

Shapes of Molecules Having Irregular Geometry

Irregular Geometry : Let us now discuss some of the examples in which regular geometries of certain molecules show distortions. Some examples have been described as follows :

1. Shape of PF_5 **molecule (AB, type)—Trigonal bipyramidal**

In PF_5 central atom is $_{15}P$ which is having the configuration $1s^2\ 2s^2\ 2p^6\ 3s^1\ 3p_x\ 3p_y^1\ 3p_z^1\ 3d^1$ in the excited state. It is able to make five bonds with the five fluorine atoms because of the presence of five unpaired electrons in its valence shell. In the formation of P F each of these electrons gets shared by five fluorine atoms. Because of this, the phosphorus atom gets surrounded by five bonded pair of electrons.

Total number of electrons present in the valence shell of P = 5

Total number of bonds made by phosphorus = 5

Total number of bond pairs of electrons around phosphorus = 5

Total number of unshared pair of electrons around phosphorus atom

$$= 1/2\ [5 - 5] = 0$$

Total number of stereoactive electron pairs present around phosphorus atom = 5 + 0 = 5

In PF_5, the central phosphorus atom has been bonded to five similar atoms by single covalent bonds and gets surrounded by only five shared pair of electrons. Therefore, PF_5, *should have regular geometry*. However, *actually* the molecule is having *irregular geometry*.

Explanation for irregular geometry : So as to attain minimum force of repulsion among five bonded electron pairs, three FPF bond angles would become 120° and two F–P–F bond angles would become 90°. The molecule gets trigonal bipyramidal geometry. In such a geometry, the central atom (P) is situated in the centre of an equilateral triangle and three P–F bonds (called equatorial bonds) get directed towards the three corners of this triangle. Among the remaining two P-F bonds (called axial bonds), one is situated above and other below the trigonal plane and also perpendicular to it.

Different bond lengths in equatorial and axial bonds : The bond length of equatorial bonds (2.04Å) has been found to be slightly less than axial bonds (2.19Å). *The reason is that, each P–F axial bond faces repulsion from three equatorial bonds at 90° and one axial bond at 180° to it while each equatorial bond faces repulsion from two equatorial bonds at 120° to it and two axial bonds at 90° to it.*

Hence, each axial bond pair (being nearer to three bond pairs at 90°) has to face greater repulsion than PF equatorial bond pairs (which are nearer to two bond pairs at 90°). In order to decrease repulsive force, axial bond pairs tend to move away from the central atom and the bond length of axial P–F bonds gets increased. Because of the difference in bond length values of axial and equitorial bonds, the PF_5 molecule is having irregular geometry.

Unstability and more reactivity of PF_5. The trigonal bipyramidal structure of PF_5 has been an unsymmetrical structure. Because of this, less stable unsymmetrical structure, PF_5 has been highly reactive as compared to SF_6 having symmetrical structure. Other examples of molecules including trigonal bipyramidal geometry include PCl_5, $SbCl_5$, etc.

Presence of One Lone Pair

1. Shape of NH_3 molecule—Pyramidal : In NH_3 the central atom is which is having the configuration $1s^2\ 2s^2\ 2p_x^1\ 2p_z^1$ in the ground state. It is able to make three bonds with three hydrogen atoms because of the presence of three unpaired electrons in it. In the formation of NH_3, each of these electrons is shared by three hydrogen atoms. Because of this, the nitrogen atom is surrounded by three shared pairs of electrons. One lone pair of electrons is still left in the valence shell of nitrogen.

Total number of electrons present in the valence shell of nitrogen=5

Number of bonds made by nitrogen = 3

Number of shared pairs of electrons around nitrogen = 3

Number of unshared pairs of electrons around nitrogen

= 1/2(5–3) = 1

Total number os stereoactive pairs present around nitrogen

= 3 + 1 = 4.

The central nitrogen atom in NH_3 gets surrounded by shared as well as unshared as well as unshared pair of electrons. Therefore, the force of repulsion among electron pairs would be unequal. Hence NH_3 molecule would be having *irregular* or *distorted tetrahedral* geometry. It is also called *pyramidal geometry* : In such a geometry, the central atom (N) is situated at the centre of a tetrahedron. The three N–H bonds get directed towards the three corners of this tetrahedron while the fourth corner gets occupied by a lone pair. Each H–N–H bond angle would be 107°.

Explanation for Decrease in Bond Angle

The valence shell of the nitrogen atom in NH_3 contains four electron pairs, as does the carbon atom in methane. Thus, NH_3 might be expected to form a regular tetrahedral configuration. However, the N–H–H bond angle in NH_3 should be 109° 28¢ but actually it is 107°. This decrease in the tetrahedral angle is due to fact that the nitrogen atom contains an unshared electron pair, called a *lone pair*.

Since the lone pair is under the influence of only the nitrogen nucleus, its electron cloud will be spread out in space to a greater extent than will be a bonding electron pair, which is under the influence of both the nitrogen and hydrogen nuclei. This greater repulsion in space of the lone pair electron will result in a greater repulsion between the lone pair and the bond pair than would result between a bond pair and another bond pair. The net result is that the lone pair causes the three N–H bond pairs to be pushed close together, thereby decreasing the bond angle form 109°, 120° to 107°.

Other examples of molecules having pyramidal structure include PH_3,PCl_3 and NF_3. However, the bond angles have been different as compared to the bond angles in case of ammonia.

2. Shape of SF_4 molecule : In SF_4 there are four bonding pairs and one lone pair. If the lone pair is not considered the expected geometry of SF4 should be tetrahedral. But the lone pair has to be considered. In such a situation the geometry should become trigonal bipyrammid. There are two positions for locating the lone pair, *i.e.*, one equatorially and other axially.

The experimental observed structure shows the position of lone pair equatorially. It causes the repelling of bonding pairs, resulting in the formation of a distorted trigonal bipyramidal structure. In this position the lone pair is under the influence of only two S–F bonds at 90°, whereas in the axial position, the lone pair is under the influence of three S–F bonds at 90°. Generally, lone pairs occupy positions where they encounter few 90° interactions.

3. Shape of TeF_5^- **:** In this molecule, there are twelve electrons in the valence shell of Te. Out of these 10 electrons form 5 bonding pairs whereas the remaining two electrons form lone pair. The expected geometry is the octahedral structure whereas the lone pair repulses the fluorine atoms to give a square pyramidal structure.

Presence of Two Lone Pairs

1. Shape of H_2O: bent or V-shaped : In H_2O the central atom is oxygen which is having the configuration $1s^2\ 2s^2\ 2p_x^{\ 2}\ 2p_y^{\ 1}\ 2p_z^{\ 1}$ in the ground state. It is able to make two bonds with two hydrogen atoms because the presence of two unpaired electrons in it. In the formation of H_2O, each of these electrons gets shared by two hydrogen atoms. Because of this the oxygen atom gets surrounded by two shared pairs of electrons. Two lone pairs of electron are still left in the valence shell.

Total number of electrons present in the valence shell of oxygen=6

Number of bonds made by oxygen = 2

Number of shared pair of electrons around oxygen = 2

Number of unshared pair of electrons around oxygen = 1/2 [6 – 2]=2

Total number of stereoactive electron pairs around oxygen in H_2O

= 2 + 2 = 4

In H_2O molecule the central oxygen atom is surrounded by shared as well as unshared pairs of electrons therefore, the force of repulsion among electron pairs becomes unequal. Hence, H_2O molecule is having an irregular or distorted tetrahedral geometry. It is also termed as *bent or angular or V-shaped* geometry. In such a geometry, the central atom (O) is situated at the centre of a tetrahedron. The two O–H bonds get directed towards the two corners of this tetrahedron whereas the two other corners get occupied by two lone pairs. The H–O–H bond angle would be 104.5°.

Explanation for decrease in bond angles : The valence shell of the oxygen atom in water contains four electron pairs and hence it should have tetrahedral structure with bond angle of 109° 28′. But the bond angle in this case is 104° 27′. The reason for this is that two of four electron pairs are lone pairs.

From the reasoning used in the çase of ammonia, one would expect an even greater repulsion between the electron clouds of the two pairs than the repulsion occurring between a lone pair and a bond pair. This greater repulsion between two lone pair results in a greater pushing together of the bond pair. Hence the H–O–H bond angle in H_2O would be expected to deviate from a tetrahedral angle to a greater extent than the H–N–H angle in NH_3. Other examples of compounds having bent geometry include H_2S, H_2Se and H_2Te. However, the bond angles have been different as compared to the bond angles in case of water.

Limitations of VSEPR Theory

(i) VSEPR theory does not explain the shapes of molecules having very polar bonds. For example, Li_2O should have the same structure as H_2O but it is linear.

(ii) VSEPR theory does not explain the shapes of molecules which have extensive delocalised π electron systems. For example, $[C(N_3)]^-$ should be pyramidal structure but it is planar.

(iii) VSEPR theory does not explain the shapes of certain compounds which have an inert pair of electrons. For example, AX_6E possessing six bonding and one non-bonding electron pairs should have distorted molecular octahedral geometry but it has regular geometry.

(iv) This theory fails to explain the shapes of certain compounds of transition elements. For example, complexes of transition metals having d^8 configuration are often square planar and not tetrahedral as predicted by Gillespie-Nyholm theory.

(v) VSEPR theory does not make any distinction between the electrons which are present in different types of orbitals, (*i.e.*, *s, p, d* or *f*) of the valence shell.

(vi) VSEPR theory fails to predict the exact bond angles in molecules such H_2O and NH_3 having distorted geometries.

Examples:

In order to understand this theory completely, one has to understand the following simple examples.

(a) Formation of H_2 molecule : When hydrogen atoms are brought sufficiently close together, it is possible that their 1s charge may interact with each other in one of the two ways.

(i) They can overlap or fuse into each other, resulting in the union of two atoms, or

(ii) They can repel each other and thus prevent any union of the two atoms.

Calculations involving applications of *Pauli's exclusion principle* reveal that the attraction between two atoms occurs when the electrons are of opposite spins, *i.e.*, when they are paired, electrons of the same spins cause repulsion. Thus, for stable bond formation, the condition is that the electrons should have the opposite spins. The concept of orbital overlap can be understood in the following manner. It has been observed that if a mole of H_2 molecule is heated to sufficiently high temperature, the reaction takes place as follows :

$$H_2(g) + 433J \rightarrow H(g) + H(g).$$

It means that, 433 kJ of energy is required to dissociated a mole of hydrogen into hydrogen atoms. In other words, the formation of a mole of hydrogen atoms gives 433 kJ of energy. Thus, the bond formation is accompained by lowering of the energy of the system. In case of the atoms where electron transfer is to favoured on energy consideration, a chemical bond may be formed in some manner which may also lead to lowering of the energy of the system.

For instance, if one or more electrons of one atom may get so close to the other atom that they may experience attractive interaction with the other nucleus (besides its own) the total energy can be lowered. A nucleus at tracts electrons in the region of apace defined as orbital. The overlapping of the orbitals of the two a toms is essential for an electron to face simulataneous attractive interactions from the nuclei of the two atoms. This overlap region should be available to electrons of both the atoms. The occupations of the overlapping orbitals by electrons may take place in the following ways.

(i) Overlapping of orbitals of the atoms which contribute one electron each. The bond formed is called covalent bond.

(ii) Overlapping of orbitals of an atom having two electrons (lone pair) by an empty orbital of an atom. The bond formed is called co-ordinate bond.

Let us now study orbital concept of covalent bonding. The overlapping of s-orbitals in the case of bonding in hydrogen. It can be seen that during overlapping, parts of atomic orbitals of the atoms occupy the same space.

The question now arises, "*how much orbital overlapping can take place?*" It is known that there are attractive and repulsive ineractions present. The repulsive interactions present. The repulsive interactions tend to increase tend to increase the potential energy whereas the attractive interactions tends to decrease the potential energy.

Considering the gradual approach of the atoms, at first the attractive interactions increase more rapidly than the repulsive interactions resulting in the decrease of the potential energy of the system. As the energy of the isolated atom is arbitrarily taken as zero, the decease in the energy of the system is shown as negative.

After a certain interatomic distance, the repulsive interactions start increasing more rapidly than the attractive interaction, thereby resulting in the increase in potential energy of the system. The position of minium energy is one where attractive and repulsive forces balance. The more the overlapping of the orbitals, the more will be the decrease in the energy of the system and hence a more stable covalent bond will be formed.

Attractive and repulsive interactions between bonding atoms : The attractive and repulsive interactions between bonding atoms are deciding factors for the formation of a bond between them.

Explanation for the Formation of H_2 Molecule

(i) *Suppose H_A and H_B atoms are at infinite distance from each other.*

In this case, the attractive and repulsive interactions between two atoms have been zero. The potential energy of isolated atoms would be arbi-traily taken as zero.

(ii) *Suppose H_A and H_B atoms are brought closder* : In this case the following attractive and repulsive forces would become operative.

(a) Attractive forces which are existing between electron of H_A and nucleus of H_B and vice versa.

(b) Repulsive forces which are existing between electrons of H_A and H_B and also between the nuclei of H_A and H_B atoms. But the increase in attractive forces has been greater than the increase in the repulsive forces, it means that energy of the system decrease.

(iii) **Suppose H_A and H_B atoms are at a distance of bond length r_0.**

In this case, three exists net attractive force between atoms. The energy of the system is minium = d and a stable H_2 molecule gets formed. Bond length for H_2 molecule = r_0 = 0.74 Å and bond energy = 436.8 kJ/mole.

(iv) **Suppose H_A ad H_B atoms are brought at a distance less than bond length :** In this case, the nuclei of H_A and H_B atoms repel each other strongly. This increase potential energy of the system and curve exhibits upward trend. The molecule is unstable.

He_2 molecule is not formed : There are *two reasons :*

(1) 1s orbital of helium (Z = 2) has been fully filled ($1s^2$). For bond formation, the overlapping orbitals should be half-filled. Thus 1s-orbital of He-atom fails to form He_2-molecule.

(2) (a) *Suppose He_A and He_B atoms are at infinite distance form each other :* In this case the attractive and repulsive interaction between two atoms would be zero. The potential energy of isolated atoms would be thus maximum.

(b) *Suppose He_A and He_B atoms are brought closer :* In this case, following attractive and repulsive interactions come into operation.

(i) *Four new attractive forces* which include two attractive forces between nucleus of He_A–atom and two electrons of He_B–atom (+) and two attractive forces of

He_B atoms and two electrons of He_A–atom *i.e.*, 4 attractive forces.

(ii) *Five new repulsive forces* : As the repulsive forces have been more than attractive forces, helium atoms fail to combine to form He_2 molecule.

Formation of Fluorine Molecule : *Fluorine* has electronic configuration as $1s^2\ 2s^2\ 2p^5$. It is difficult to consider its bonding by one orbital similar to that in the hydrogen molecule. The 2p orbitals are split into three sub orbitals under the influence of a magnetic field. These five p electrons are placed, two each in two of p-orbitals and one in the remaining p orbital. The two atoms of fluorine. The two p-orbitals with one electron in each can then form a single bonding orbital with two electrons. It is similar to the hydrogen molecule. In the figure, only p-orbitals are shown.

Formation of Hydrogen Fluoride : Let us consider the hydrogen fluoride molecule, HF. There are five electrons in the 2p orbital of fluorine, *i.e.*, two paired and one unpaired. If the orbital containing the unpaired electrons overlaps the 1s orbital of the hydrogen atom, the spins of the two electrons are opposite. Then just as in the case of hydrogen molecule, the two waves knit together establishing a bond between two atoms.

Successes of the Theory : *The Heitler-London* theory is very useful because many cases of covalent bond formation can be treated in terms of it.

Objectives : Various objections are.

(a) **In the Heitler-London Theory :** it is assumed that the electrons in a shared pair come form different atoms. Thus, the existence of coordinate linkage is ruled out in which electrons come from one atom only and not from two.

(b) This theory fails to explain the structure of perbromates.

(c) The theory fails to explain the structure of odd electron molecules.

(d) In the **Heitler-London Theory :** it is assumed that such paired electrons may be involved in they can be unpaired without the expenditure of excessive energy. In the formation of PF_5, electrons are to be unpaired with required energy. Thus it violates this theory.

(e) It fails to account for the *directional characteristics of covalent jonds*

Reasons for the Failure of Heitler-London Theory : Let us consider two isolated hydrogen atoms. These can be described by the wave functions Ψ_A and Ψ_B, where A and B represent the two hydrogen atoms. If these two atoms are quite far apart, these is no interaction and the wave function for this system of two hydrogen atoms may be given as

$$\Psi = \Psi_A (1) \Psi_B (2) \qquad ...(1)$$

where 1 is the electron of atom A and 2 is the electron of atom B.

If we bring the two hydrogen atoms together to form a molecule, they will affect one another and the individual wave function will undergo change. The bonding electrons must be of opposite spin. If they are of parallel spin, no bonding occurs.

By using the wave function, Ψ, as given in equation (1), one can calculate the energy of the system as the two atoms are gradually brought together. Then the energy curve I having a minium of 0.25 eV at an internuclear distance of about 0.9Å is obtained. But the actual experimental bond energy for the hydrogen molecule is 4.7 eV which is much is much higher than this.

Now one may ask questions about the differences between the experimental values and the values obtained from solving equation (1) These differences arise due to the simple fact that equation (1) does not described the hydrogen molecule correctly. A number of limitations have been placed on equation (1). These restrictions were applied by *Pauling and Slater* in 1931. These restrictions are as follows.

(1) First Improvement (Exchange Energy) : As electrons are similar, they are not distinguishable from each other. Hence, it is not proper to lable the electrons as 1 and 2. Even if we are able to distinguish and identify the electrons, we will not be sure that electron 1 was always on atom A whereas electron 2 was always on atom B. Therefore, equation (1) needs modification and it is possible to show the probability that the electrons could exchange their positions so that the electron 1 at any time is associated with atom B and *vice versa.* Mathematically, this situation can be introduced in equation (1) by incorporating a second term *i.e.*,

$$\Psi = \Psi_A (1) \Psi_B (2) + \Psi_A (2) \Psi_B (1) \qquad ...(2)$$

When equation (2) is solved for the energy associated with it, curve II is obtained which gives bonding energy of 3.14 eV and bond distance of 0.689 Å. The bonding energy, as shown by curve II, compares well with the experimental curve E. The additional lowering in energy has been obtained by virtue of exchange of electrons between the two hydrogen atoms and is known as *exchange energy*. This exchange energy shows that the molecule is more stable and it can also be explained by assuming that now the electron has a larger volume to move.

(ii) Second Improvement (Shielding Effect) : We know that the electrons possess the capacity to shield one another. It means that the effective atomic number Z* is generally less than Z, the atomic number. If one considers the shielding factor, *i.e.*, in which one electron shields the other from the attractive nuclear force, the wave functions Ψ_A and Ψ_B need further changes. When these are considered, the curve III is obtained which approaches the experimental value of curve IV, to a greater extent. The values for the curve III are 3.78 eV for bonding energy and 0.,743 Å for bond distance.

(iii) Third Improvement (Ionicity) : According to equation (2), electrons can exchange places. But exchange takes place simultaneously. It means that if electron 1 is associated with atom B, the other electron has also gone over to A. But this situation is not correct though electron repulsion would tend to keep the atoms apart. Suppose two electrons were present on one atom. It means that the atom would acquire a negative charge and other hydrogen atom would acquire a positive charge.

Thus, there are two ionic structures viz.,

$$H+ - H^- \leftrightarrow H^- H^+$$

These ionic equation have to be added to the s structure as represented by equation (2) which represents a covalent structure, H–H. Thus :

$$H - H \leftrightarrow H^+ H^- \leftrightarrow H^- H^+$$

Therefore, equation (2) becomes as follows.

$$\Psi = \Psi_{A(1)} \Psi_{B(2)} + \Psi_{A(2)} + \Psi_{B(1)} + \lambda \Psi_{A(1)} \Psi_{A(2)} + \lambda \Psi_{B(1)} \Psi_{B(2)} \quad ...(3)$$

In the above equation, the first two terms represent the covalent structure whereas the remaining two structures represent the ionic structure.

2. **Pauling Slater Theory :** Now we are in a position to summarise the postulates of valence bond theory by taking in to account the above facts. Pauling and Slater developed Heitler-London theory to account for the *directional characteristics in covalent bonds.* To take the geometrical s-apes of orbitals into account and to relate structural chemistry to the covalent bond, the Pauling–Slater's theory suggests that :

   (a) The atoms which unite to form a molecule completely retain their identities in the resulting molecule.

   (b) A covalent molecule is formed due to the overlapping of atomic orbitals. If there are two atoms each having unpaired electron, the atomic orbitals having unpaired electrons overlap resulting in the formation of a covalent bond between these two atoms in which the spins of two electrons get mutually neutralised. If these atomic orbitals have parallel spins, repulsion will take place and no molecule will, therefore, be formed.

   (c) If the atomic orbitals are having more than one unpaired electron, multiple bond formation is possible. In nitrogen molecule, for example, three bonds are possible because N atom has three unpaired electrons.

   (d) The number of covalent bonds formed by an atom is equal to the number of unpaired electrons it has.

   (e) *The strength of the covalent bond is proportional to the extent of the overlapping of the two atomic orbitals.* The greater the overlap between the atomic orbitals, the greater is the strength of the resulting covalent bond. The extent of overlap of two atomic orbitals of wave function Ψ_A and Ψ_B is represented by overlap integral, S, which is as follows.

$$S = \int \Psi_A \ \Psi_B \ dr$$

From the above relation, it is evident that :

(i) If S is zero, there is no overlap, *i.e.*, there is neither attraction nor repulsion between the combining atoms. Thus, no bond formation is possible.

(ii) If S is positive, there is an increase in the electron density between two nuclei which results in an increase in attraction between the two combining atoms, forming a covalent bond.

(iii) It S is negative, there occurs a decrease in electron density between the nuclei for two combining atoms. Thus, the repulsion between them is increased and hence no bond formation is possible.

(f) *Overlapping takes place between orbitals of only those electrons, which take part is bond formation and not between other orbitals of the atoms.*

(g) *Between two orbitals of the same energy content, the overlapping axially will form a stranger bond.*

(h) *A spherically symmetrical orbital, i.e., s-orbital does not show any directional preference whereas non-spherical orbitals, viz, p, d, f etc., tend to form a bond in the direction of maximum electron density within the orbital, i.e., along their axes.* For example,

(i) As the charge density of s orbital is spherically symmetrical, the two s-orbitals will overlap to the same extent in all directions.

(ii) The three p-orbitals lie on mutually perpendicular axes having nucleus at the centre. Whenever p-orbitals overlap, they tend to form a stronger bond in the direction of their maximum electron density.

(iii) When a s-orbital overlaps p-orbital, the bond is formed but it is weaker than p-p bond but stronger than s-s bond.

Types Of Overlapping

Different types of overlapping, which we come across, are :

(a) **s-s overlapping :** This type of overlapping is possible between atoms having half-filled s-orbitals in their outermost energy shells. In the formation of a molecule of hydrogen spherical 1s orbital of one atom overlaps with that of the other forming a single covalent bond.

In the LiH molecule, Li–H bond is formed due to the overlapping of 1s orbital of hydrogen with 2s orbital of lithium.

(b) **s-p overlapping :** This type of overlapping occurs when s-orbital of one atom overlaps with p-orbital of another. For example in the formation of H–F molecule fluorine has atomic number 9 ad its electronic configuration is written as $1s_2$ $2s^2$

$2p^5$. Further p-orbital consists of three sub–orbitals along the three axes, p_x, p_y and p_z,. The complete electronic configuration of fluorine atom is $1s^2\ 2s^2\,2p_z^2\ 2p_y^2\ \ 2p_x^2$.

Thus one of the orbital has an unpaired electron available for bond formation. The 1s orbital of hydrogen then overlaps with 2ps orbital of fluorine to form a molecule of HF. Similarly the molecule of lithium fluoride (Li–F) is produced by the overlapping of 2s orbitals of lithium with 2px orbital of fluorine.

(c) p-p overlapping : It is illustrated by the formation of fluorine molecule. The electronic configuration of fluorine is $1s^2$, $2s^2$, $2p_z^2, 2p_y^2\ \ 2p_x^2$; there is one half-filled 2px orbital. When the overlapping of these two orbitals takes place along their nuclear axis, a sigma bond is formed, as shown below : Similarly, the formation of Cl–Cl and Br–Br can be explained. Examples. From the concept of valence bond theory, the structure of water (H_2O) and ammonia (NH_3) molecules can be explained.

(a) Structure of NH_3 molecule : The electronic configuration of nitrogen is :

$$1s^2\ 2s^2\,2p_x^1\ 2p_y^1\ \ 2p_z^1.$$

As the nitrogen atom has three unpaired electrons, it is for the reason that it exhibits a valency of three.

When nitrogen atom combines with three hydrogen atoms to form a molecule NH_3 the three N–H bonds should be at right angles to one another. Thus, the molecule of NH_3 shows a pyramidal structure. Actually the angle between N–H bonds in NH_3 is 108° instead of 90°. This can be explained from the *polar character* of the bond. As nitrogen has more affinity for electrons than hydrogen, the former (nitrogen) acquires a little negative charge (say, $-\delta$) and the latter (H) a little positive charge (say, $+\ \delta$). The positively charged hydrogen atoms repel each other and get further apart than 90°.

(b) Structure of H_2O molecule : The electronic configuration of exygen atom is given as :

$$1s_2\ 2s^2\,2p_x^2\ 2p_y^2\ \ 2p_z^2$$

As oxygen atom has two paired electrons, it means that it has a valency of two. In water molecule (H_2O), the oxygen atom combines with two hydrogen atoms. The two O–H bonds in the formation of water should be at right angles to each other. Actually the angle between two O–H bonds in H_2O is 105° instead of 90°. Again, it can be explained on the basis of *polar character of the bond.* Due to the great affinity of electrons for oxygen atom, it results in the accumulation of slight positive charge on H atom and a slight – ve charge on the oxygen molecule repel each other and get further apart than 90°.

An alternative explanation is offered by the concept of *hybridisation.*

Limitations of Valence Bond Method : These are.

(i) It did not explain the formation of molecules such as PCl_5, SF_6 and IF_7 in which the central atom is having 10, 12 and 14 electrons respectively instead of an octet of electrons.

(ii) It did not explain the bonding in electron dificient molecules like B_2H_6; in these molecules, the central atom possesses less electrons than the required octet of electrons.

(iii) It did not explain the existence of paramagnetic molecules like NO, NO_2 and ClO_2,

(iv) It failed to explain the nature of attractive forces that would become operative when a covalent bonds is being formed.

(v) It did not predict the amount of energy which would be involved during the formation of a covalent bond.

(vi) It failed to predict the geometry of covalent molecules like H_2O, CH_4, etc.

(vii) This theory only explains the formation of covalent bond in which the shared pair of electron s comes form two different atoms. However, it offers no explanation for the formation of a co-ordinate bond in which one of the bonded atoms furnishes both the electrons.

(viii) By this method the structure of exygen molecule should be diamagnetic, *i.e.*, all the electrons present on it should be paired electrons. Experimentally, the oxygen molecule has been found to be paramagnetic, indicating the presence of unpaired electrons in oxygen molecule.

(ix) This theory fails to explain the formation of odd electron molecules or ions like H_2^+ ion in which no pairing of electrons occurs.

(x) For any particular molecule or ion, one has to write several canonical structures and even then one is not sure about the true structure which is only a resonance hybrid of these structures. Thus, it is not possible to write one structure which may offer explanation for all the structural features of that molecule or ion.

(xi) It fails to explain the bonding in metals and intermetallic compounds and also in electron deficient molecules.

(xii) It fails to explain the presence of partial ionic nature of a covalent bond.

Inspite of these limitations, the valence bond theory finds wide applications,. It is generally used as a qualitative guide in determining geometry of molecules.

MOLECULAR ORBITAL THEORY

Why molecular orbital theory? In the valence bond method following assumptions are to be made :

(i) Electrons in molecules are localised as if they are in isolated atoms. It means that electron orbital of an isolated atom is not changed if a nucleus of another atom is brought near to it.

(ii) Electrons retain their characteristics even in the presence of other nuclei.

(iii) The wave function of the molecule is the product of the atomic wave functions of the individual atoms.

The above three postulates are not correct and are product of the cases of over simplification. These are taken into account in molecular orbital theory which is therefore more correct.

Valence Bond and Molecular Bond Methods

The valence bond method. mainly due to the work of Heitler, *London, Slater* and *Pauling.* considers the molecule as being made up of atoms with electrons in atomic orbitals on each atom. *Thus, a molecule is treated as if it were composed of atoms which, to some extent, retain their individual character when linked to other atoms.*

The molecular orbital method– mainly due to the work of *Hund, Jones* and *Mulliken– treats a molecule in the same way as an atom, except that in the molecule an electron moves in the field of more than one nucleus, i.e., molecular orbitals are polycentric.* An over simplified pictorial representation of the two methods in illustrate.

In A and are two isolated atoms. In A and B combine in accordance with valence bond method, *i.e.*, atoms A and B do not lose their individual character while forming a molecule AB.

In atoms A and B combine to form molecule AB according to the molecular orbital method. In this case two electrons move in a field of two nuclei, *i.e.*, molecular orbital are polycentric. Thus, in molecular orbital method, atoms A and B lose their individual character.

The *basic principles of molecular orbital theory* are given as follows:

1. When nuclei of two atoms come close to each other, *their atomic orbitals interact resulting in the formation of molecular orbitals*. The atomic orbitals of the atoms in a molecule completely lose their identity after the formation of molecular orbitals.
2. Each molecular orbital may be described by a wave function Ψ, known as molecular orbital wave function.
3. Ψ is such that Ψ^2 represents the probability density of electron density.
4. Each molecular orbital wave function (Ψ) is associated with a set of quantum numbers which ascertain the energy and the shape of the molecular orbital.
5. Each Ψ is associated with a definite energy value and the total energy of the molecules is the sum of the energies of the occupied molecular orbital.
6. Electrons tend to fill the molecular orbitals in the same way as they fill the atomic orbitals following the Aufbau principle, Pauli's exclusion principle and the Hund's rule of maximum multiplicity.
7. Each electron in a molecular orbital belongs to all the nuclei present in the molecule.
8. Each electron moving in a molecular orbital is having a spin of $+\frac{1}{2}$ or $-\frac{1}{2}$.

The **basic difference** *between an atomic orbital and a molecular orbital* is that an electron in an atomic orbital is influenced by *one positive nucleus only* while an electron in molecular orbital is influenced by two or more nuclei depending upon the number of atoms contained in a molecule.

There are two methods for getting an approximate yet at the same time a satisfactory idea about molecular orbitals.

1. Linear combination of atomic orbital method (LCAO) method.
2. United atom method.

Let us discuss these one by one.

1. **LCAO Molecular Orbital Theory :** The molecular orbitals can be obtained by the method of linear combination of atomic orbitals (LCAO).

In LCAO method two nuclei must approach one another along a line When they come close to one another, two atomic orbitals (one form each bonded atom) having comparable energies and possessing large overlap combine to form two molecular orbitals. One of the new molecular orbitals has lower energy than the combining atomic orbitals and the other has higher energy than the combining atomic orbitals.

The molecular orbital with higher energy than the atomic orbitals gives rise to a repulsive state and is called *antibonding orbital* while the molecular orbital with lower energy than atomic orbitals gives rise to an attractive state and is called the *bonding orbital.* It may be further pointed out, that electrons placed in the antibonding orbital tend to increase the energy of molecule, thus decrease its stability and the electrons placed in bonding orbitals tend to decrease the energy of the system, and thus increase its stability.

The bonding and antibonding orbitals are distinguished from each other by putting an asterisk (*) on the antibonding orbitals.

Let us illustrate the formation of molecular orbitals in hydrogen molecule. Before we discuss this theory, we will attempt to explain the meaning of an electron wave (Ψ).

The electron wave above the imaginary horizontal line is given a + ve sign while that below a line is –ve. One should not confuse these sign with the signs of charged bodies. These simply depend upon the value of co-ordinates. In the formation of hydrogen molecule, s orbital of one hydrogen atom overlaps 1s orbital of another hydrogen atom to form two molecular orbitals. Recalling the wave nature of electrons there are two ways of combining atomic orbitals, (*i.e.*, electron waves).

(i) One way is by addition.

The two dots represent the two nuclei. In this case both the electron waves are on the same side (+ ve side) and on adding result in the formation of a molecular orbital of the type as shown on the right hand side. In such a molecular orbital, the region between the two positive nuclei where 1s orbitals overlap will have high probability. The electron density is illustrated. As the electron density between the two nuclei is maximum, it shields the nuclei from mutual repulsion and holds them at a certain distance from each other which is actually equal to the bond length.

Such a bonding orbital is called a bonding molecular orbital and by convention it is designated as σ (1s) orbital. The sign σ signifies that the orbital is symmetrical about the molecular axis joining the two nuclei and the script 1s indicates that molecular orbital is formed by the combination of two 1s atomic orbitals.

(ii) The second way of combining the two orbitals is by substraction.

Here the two electron waves are oriented in two opposite directions. Therefore, they cancel out each other and thus the probability of finding electrons in the region of overlap is practically nil. As there are no electrons in this region, there is not shielding from nuclear repulsion. Hence there is no bonding and the nuclei are repelled from each. Such a molecular orbital is known as antibonding molecular orbital and is designated as σ^* (1s). the sign * (asterisk) signifies the antibonding nature of molecular orbital.

A bonding molecular orbital (BMO) : *on account of minimising of repulsion between the nuclei, is associated with much lower energy than either of the atomic orbitals from which it is derived.* An *anti-bonding molecular orbital (ABMO), on the other hand, on account of greater repulsion between the nuclei, is associated with higher energy than the uncombined AOs.*

(a) Features of Bonding and Antibonding Molecular Orbitals : Suppose Ψ_A and Ψ_B are the wave functions of two atoms A and B respectively. If the lobes of the combining atomic orbitals with the same sign overlap, the bonding molecular orbital Ψ_b, is formed..

$$\Psi_b = \Psi_A + \Psi_B \quad ...(1)$$

When the lobes of the combining atomic orbitals of opposite signs overlap the new molecular orbital is obtained. It is denoted by Ψ_a, the antibonding orbital.

$$\Psi_a = \Psi_A - \Psi_B \qquad ...(2)$$

The electron distribution in a molecular orbital can be obtained by squaring the wave functions. On squaring equations (1) and (2), we get for bonding and antibonding orbitals such as

$$\Psi_b^2 = \Psi_A^2 + \Psi_B^2 + 2\Psi_A \Psi_B \qquad ...(3)$$

and

$$\Psi_a^2 = \Psi_A^2 + \Psi_B^2 - 2\Psi_A \Psi_B \qquad ...(4)$$

The equation (3) and (4) represent the probability functions for bonding and antibonding molecular orbitals. These two differ by the cross term $2\Psi_A \Psi_B$. This integral $\int \Psi_A \Psi_B$ is the known as the overlap integral, S. The value of S is numerically very small and, therefore, is neglected in calculating the wave functions of bonding and antibonding orbitals. When the overlap integral S is completely neglected, the molecular wave functions become as follows :

(i) For bonding orbitals,

$$\Psi_b = \frac{1}{\sqrt{(2)}} (\Psi_A + \Psi_B)$$

(ii) For antibonding orbitals,

$$Y_a = \frac{1}{\sqrt{(2)}} (Y_A - Y_B).$$

(b) One can make the clear distinction between bonding and antibonding molecular orbitals by showing the *three-dimensional contours* of electron charge densities or probability densities (viz. Ψ^2) around the two nuclei A and B for these orbitals, as illustrated (a) and (b).

In a bonding MO, the high electron charge density contours are localised on the two nuclei but tend to polarise *towards the space* located in between the two nuclei. The low electron charge density contours are, however, spread over both the nuclei. This implies enhancement of electronic charge in space in between the two nuclei. The electron density in the bonding orbital tends to move in the region around the two nuclei. Due to the high electron density, the nuclei are shielded from each other and there is greater attraction of the nuclei for the electrons. This results in a lowering of energy and is therefore a bonding structure.

In an antibonding MO, the high as well as low electron charge density contours are localised on the two nuclei and tend to polarise *away from the space* falling in between the two nuclei. This implies depletion of electronic charge from the space in between the two nuclei. In an antibonding orbital the electrons are in those regions in space where they are not under the attraction of both the nuclei. *The probability of finding the electron in between the nuclei is negligible. The electron density is concentrated on any one nucleus at a time. Every electron in an antibonding molecular orbital contributes to repulsion of the two atoms because it possesses higher energy than the atomic orbitals from which it is formed.*

ENERGY LEVELS OF MOLECULAR ORBITALS

In the method of molecular orbitals, the *s, p* and *d* orbitals in a simple atom are replaced by σ, π and δ orbitals in a molecule. Let us consider the combination of various orbitals to form molecular orbitals:

(i) **Combination of s-orbitals :** A combination of two similar 1s–atomic orbitals will yield two molecular orbitals to accommodate the available electrons. One of the molecular orbitals is a bonding orbital (σ 1 *s*) and the other is antibonding (σ* 1 *s*).

Combination of two 2s atomic orbitals will produce a *2s* and a* *2s* molecular orbitals. These will be of higher energy than o Is and O* Is orbitals since they (σ 2s and σ* 2s) are formed from atomic orbitals of higher energy.

Combination of p-Orbitals : p-orbitals can overlap, and be combined either end to end, to give σ-molecular orbitals or broad side on to give π-molecular orbitals. Let us consider the various p-orbitals.

(ii) **Combination of $2p_x$ Orbitals :** Two $2p_z$, atomic orbitals overlapping end to end can form a bonding molecular orbital as well as antibonding molecular orbital. Both the bonding and antibonding orbitals are symmetrical about bond axis, *i.e.*, x-axis. These are, therefore , called σ-bonds. These two are denoted by σ $2p_z$ and σ* $2p_z$.

(iii) **Combination of $2p_x$ or $2p_y$ Orbitals :** If two $2p_x$ orbitals be considered, they can overlap side to side and not end to end as in the case of $2p_z$ orbitals. The resultant molecular orbitals will therefore, he called **p** orbitals. The formation of bonding orbital, designated as π_x, 2p while the corresponding non-bonding orbital designated as σ^* 2p.

Similarly, the formation of p, 2p and π_y^* 2p from the combination of two 2p, orbitals can be explained. The only difference is that the atomic and molecular orbitals lie now perpendicular to the plane of the paper.

Table 7 : Notations used to designate the various molecular orbitals are shown below

Molecular orbitals Bonding	Antibonding	*Atomic orbitals from which the molecular orbitals are formed*
σ 1*s*	σ^*1s	1*s* orbitals of the two atoms
σ 2s	σ^* 2s	2s orbitals of the two atoms
σ 2s	σ^*2p	$2p_z$ orbitals of the two atoms
$\pi_x 2p$	σ_x^* 2p	2*px* orbitals of the two atoms
$\pi_y p$	σ_y^* 2p	2*py* orbitals of the two atoms

Differences between bonding and antibonding molecular orbitals: The differences between bonding and antibonding molecular orbitals may be summed up as follows :

Table 8

	Bonding MO	*Antibonding MO*
1.	A bonding MO gets formed by the addition-overlap of atomic orbitals.	An antibonding MO gets formed by the subtraction-overlap of atomic orbitals.
2.	It is having lower energy than the AOs from which it is formed.	It is having higher energy than the AOs from which it is formed.
3.	The electron charge density in between the nuclei is high and hence the repulsion between the nuclei is very low. This causes stabilisation of the bonding MO. In other words, the electrons in the bonding MO tend to favour stable bond formation.	The electron charge density in between the nuclei is very law and hence the repulsion between thenuclei is high. This causes de-stabilisation of the anti-bonding MO. In other words, the electrons in the antibonding MO tend to oppose bond formation.

Conditions for the formation of atomic orbitals : It is important to understand that any two atomic orbitals may not combine to form molecular orbitals may not combine to form molecular orbitals. In fact there are certain limitations to the combination of atomic orbitals which are summarised below :

(i) The energies of atomic orbitals must be similar in magnitude. According to this rule, a homonuclear ditomic molecule will not be formed if 1s-orbital of one atom overlaps with 2s-orbital of another atom. Similarly, a molecule will not be formed if 2s orbital of one atom overlaps with 2p-orbital of another atom because 1s, 2s and 2p-orbitals possess different energies. But in the case of formation of heteronuclear molecules, such combinations may be possible.

(ii) The atomic orbitals will combine only if they overlap to a considerable extent. The greater the overlap of atomic orbitals, the greater is the build-up of the charge between the nuclei. This rule is often referred to as the principle of maximum, overlap.

(iii) The atomic orbitals must have the same symmetry about the molecular axis. For example, p_x–orbital of one atom will not overlap with a s-orbital of another atom because the symmetry of s-orbital is not the same as that of p-orbital. In fact, the ++ overlap is cancelled by +– overlap. A p_z–orbital however, will combine with a s-orbital since both have the same symmetry.

A p_y–orbital will not combine with s-orbital as they possess different symmetry.

Electronic Configurations of Homonuclear Molecules

With the help of the various principles discussed above, the electronic configurations f same of the common molecules may be written as below.

1. Hydrogen molecule : When 1s orbitals from two hydrogen atoms overlap, σ 1s and σ^* 1s molecular orbitals are formed. Since each hydrogen atom has only one electron, both electrons in the hydrogen molecule will be in the s 1s bonding orbitals, the antibonding orbital will remain vacant. The electronic configuration of the H_2 molecule is, therefore, σ $1s^2$.

Number of covalent bonds in the molecule will be 1/2 (2 – 0) = 1. In other words, the two hydrogen atoms in the H2 molecule are bonded by a single covalent bond. In bonding molecular orbital σ 1s, there is an increase in electron charge density in the region between the two nuclei in hydrogen atoms. Thus, there is an accumulated negative charge between the two nuclei which holds these together at an equilibrium internuclear distance.

In antibonding molecular orbital, σ* 1s, there occurs a decrease in the charge density in the region between the two nuclei of hydrogen atoms. In the case, the negative charge is withdrawn from the region between the hydrogen nuclei and this increase the repulsive forces between the nuclei.

In this region, the electrons repel each other. Thus, one can say that in antibonding molecular orbital, an electron spends half of its time on each of the individual hydrogen atom and consequently is never found in the region between the two hydrogen atoms. It means that the probability of finding the electron in the region between two hydrogen atoms in antibonding orbital is decreased.

In antibonding molecular orbital, there is a line between the two probability contours of hydrogen atoms. This is called a nodal plane, and at this nodal plane the electron charge density is zero.

2. Hydrogen molecule ion, H_2^+ : This molecule has been detected spectroscopically when electric discharge is allowed to pass through hydrogen gas under reduced pressure. The molecule has only one electron and its electronic configuration is H_2^+ : $\sigma\,(1s)^1$ and *bond order* = 1/2 (1 – 0) = 1/2. The molecular orbital energy level diagram for H_2^+ ion.

The positive value of bond order reveals that the bond does get formed and the molecule is stable. However, the bond in H_2^+ ion is weaker than that in H_2 molecule. The bond dissociation energy of H_2^+ ion (269 kJ mol–1) has been some what less than that of H_2 molecule (458 kJ mole^{-1}). The bond length in H_2^+ ion (1.04 Å) has been larger than that in H2 molecule (0.74 Å). These observations support the fact that bond in H_2^+ is weaker than the bond in H2 molecule.

3. Helium molecule, He_2 : Each helium atom is having two 1s–electrons and thus there are four electrons to get accommodated in He_2 molecule. As each molecular orbital, just like atomic orbital, can accommodate only two electrons of opposite spins, only one pair can be accommodated in the bonding s (1s) molecular orbital. The other pair will have to be accommodated in the antibonding σ* (1s) MO.

The electronic configuration He_2 is, thus, He2 : σ $(1s)^2 \overset{*}{\sigma}$ (1s)

The zero bond order reveals that the molecule does not exist. Calculations have shown that *the antibonding effect of an antibonding MO is stronger than the bonding effect of a bonding MO. Consequently, there is no possibility whatsoever for the existence of He_2 molecule.* Thus, helium is monomatomic.

4. Lithium molecule : Each lithium atom has an electronic configuration of $1s^2\ 2s^1$. We have a total six electrons from the two lithium atoms which are to be filled. Only the 2s electrons (valency electrons) are involved whereas the low lying $1s^2$ electrons do not take part in bonding. The molecular orbital energy diagram.

Its molecular orbital configuration is $[(\sigma\ 1s)^2\ (\sigma^*\ 1s)^2]\ [(\sigma\ 2s)^2]$.

Bond order = 1/2 [4 – 2] = 1 (one s bond). From the bond order, it is evident that Li_2 molecule should be formed. In fact, Li_2 species are present in sufficient concentrations at the boiling temperature of lithium.

Hence there is one Li – Li sigma bond. The bond dissociation energy of LiLi is appreciably low *i.e.*, 105kJ mol^{-1}. The bond length of the Li–Li is 2.67Å.

Comparison of Li–Li *and* H–H *bonds* :

(i) The sigma bond in Li–Li molecule is some what weaker than the sigma bond in H_2.

(ii) In lithium molecule, the outer electrons in σ(2s) molecular orbital are shielded by the inner electrons. Hence the attraction between the nuclei and the electrons in σ(2s) molecular orbital is less.

5. Beryllium molecule, Be_2 : Each of the beryllium atom possesses electronic configuration $1s^2\ 2s^2$. So there is a total of eight electrons to be filled in four molecular orbitals. The configuration is $[\sigma\ 1s)^2\ (\sigma^*\ 1s)^2]$ $[(\sigma\ 2s)^2]$.

The bond order is 1/2 (4 – 4) = 0.

This indicates that molecular beryllium will not exist. This is in accordance with the experimental results.

6. Boron Molecule, B_2 : The electronic configuration of each boron atom is $1s^2\ 2s^2\ 2p^1$. It means that the outer valency shell contains three electrons. When two boron atoms combine to form the molecule, B_2, six electrons will be accommodated in the various molecular orbitals of B_2.

Out of the six electrons, there are four electrons which are present in two 2s orbitals. Two of these will get accommodated in the σ^* 2s orbital which has the lowest energy and the other two in σ^* 2s orbital.

Now, we are left with two π electrons. According to energy level diagram, both these electrons must occupy the empty σ 2p orbital of lower energy. If it is so, the molecule B_2 should be diamagnetic. But actually it is paramagnetic.

In order to explain this anomaly, it is suggested that two electrons of the two p-orbitals of two respective boron atoms will occupy π_x, 2p and p_z 2p. Each orbital will have one electron. As both these orbitals are having one unpaired electron each, the molecule B_2 is expected to be paramagnetic.

As both $\pi_x{}^2p$ and $\pi_y{}^2p$ are bonding orbitals, the B2 molecule is expected to be stable. The structure of B2 molecule is as follows :

$$(\sigma\ 1s)^2\ (\sigma^*\ 1s)^2]\ [(\sigma\ 2s)^2\ (\sigma^*\ 2s)^2\ (\pi_x\ 2p)^1\ (\pi_y\ 2p)^1].$$

The bond order = 1/2 (6 – 4) = 1.

The letters KK denoted the closed K shell structure. It is assumed that these closed shells do not enter into bonding.

The molecule has only one bond.

The bond dissociation energy of B_2 molecule has been found to be 289 KJ mol^{-1} and bond length equal to 1.59Å. The molecular orbital energy level diagram for B_2 molecule is presented.

7. Carbon molecule : The electronic configuration of carbon is $1s^2\ 2s^2\ p^2$. The outer shell of each carbon atom has four electrons.

Thus, there are (4 + 4) or eight electrons which are to be accommodated in the various molecular orbitals of C_2. There are four x-electrons. Two of these will be accommodated in (σ 2s) orbital and remaining two in the (σ^* 2s) orbital (Fig. 5.55). We are now left with four p-electrons which are to be accommodated. Both the (π_x 2p) and (π_y 2p) being of lowest energy, will get completely accommodated by taking a pair of electrons each. The structure of C_2 is.

$$[(\sigma\ 1s)^2\ (\sigma^*\ 1s)^2]\ [(\sigma\ 2s)^2\ (\sigma^*\ 2s)^2\ (\pi_x\ 2p)^2\ (\pi_y\ 2p)^2]$$

From the above, C_2 molecule does not have any unpaired electrons. It means that it is diamagnetic. Actually, it is so.

Bond order = 1/2 (8 – 4) = 2

A bond order of two is expected which is in accordance with the observed bond energy of 627.9 mole. The bond dissociation energy of C_2 molecule has been found to be 627.9 mol^{-1} and bond length equal to 1.31Å.

8. Nitrogen molecule : Electronic configuration of the nitrogen atom is 1s2 2s2 $2p_x^1\ 2p_y^1$. When the two atomic orbitals from two nitrogen atoms combine, there will be 4 electrons in the N_2 molecule and these will be distributed themselves in the various molecular orbitals as follows.

$$[(\sigma\ 1s)^2\ (\sigma^*\ 1s)]\ [(\sigma\ 2s)^2\ (\sigma^*\ 2s)^2\ (\sigma\ 2p)^2\ (\pi_x\ 2p)^2\ (\sigma_y\ 2p)^2].$$

The three antibonding orbitals biz. $\pi_x^* 2p$, $\pi_y^* 2p$ and $\sigma^* 2p$ remain vacant, σ 1s and σ* 1s orbitals do not take part in bond formation and are known as **non-bonding orbitals.**

$(\sigma\ 2s)^2$, $(\sigma\ 2p)^2$, $(\pi_x\ 2p)^2$ and $(py\ 2p)^2$ from bonding orbitals while (s* 2s)2 forms antibonding orbitals. Therefore, number of bonding electron is 8 while the number of antibonding electrons is 2 only. Thus, the number of covalent bonds is 1/2 (8 – 2) = 3. This means that two nitrogen atoms in the N_2 molecule are bonded by a triple covalent bond. The formation of the nitrogen molecule.

In the formation of N_2 molecule only one antibonding orbital is involved while the remaining four are all bonding orbitals. Hence the nitrogen molecule is highly stable. This is confirmed by its high bond dissociation energy viz., 945.6 kJ mol^{-1} and small bond length equal to 1.10Å. Also N_2 molecule is diamagnetic because there are no unpaired electrons in the molecule.

9. Oxygen molecule : The atomic number of oxygen atoms is eight. The electronic configuration of the oxygen atoms is $1s^2\ 2s^2\ 2p_x^2\ p_y^2$ $2p_z^2$. When the two oxygen atoms combine, there will be 16 electrons in the O_2 molecule and these will be distributed in the various molecular orbitals as follows :

$[(\sigma\ 1s)^2\ (\sigma^*\ 1s)^2]\ [(\sigma\ 2s)^2\ (\sigma^*\ 2s)^2\ [\sigma\ 2p)^2\ (\pi_x\ 2p)^2\ (\sigma_y\ 2p)^2\ (\pi_x^*\ 2p)^1$ $(\pi_y^*\ 2p)^1$. $[(\sigma\ 1s)^2$ and $(\sigma^*\ 1s)^2]$ do not take part in the chemical bonding because orbitals containing valency electrons take part only.

$(\sigma\ 2s)^2\ (\sigma\ 2p)^2\ (\pi_x\ 2\ p)^2\ (p_y\ 2p)^2$ are bonding orbitals and, thus, they contribute eight electrons. $(\sigma^*\ 2s)^2, (\pi_x^*\ 2p)^1$ and $(\pi_y^*\ 2p)^1$ are antibonding orbitals and they contribute 2 + 1 + 1, *i.e.*, 4 electrons.

Thus, the number of covalent bonds will be 1/2 (8 – 4) = 2, *i.e.*, the two oxygen atoms in O_2 molecule are bonded by a double bond. The oxygen molecule is represented. Due to the presence of four electrons in anitbonding MOs oxygen molecule is less stable than nitrogen molecule. Its bond dissociation energy is much less *i.e.*, 494.6 kJmol^{-1}, and bond length is larger *i.e.*, 1.21Å.

10. O_2^- –ion : This type of ion, also called the superoxide ion is present in potassium superoxide, KO_2. This ion has seventeen electrons. The structure of this ion is related to that of the oxygen (O_2) molecule. The ion possesses one extra electron which occupies either $\pi_x^*\ 2p$ or $\pi_y^*\ 2p$. Thus,

$[(\sigma\ 1s)^2\ (\sigma^*\ 1s)^2]\ [(\sigma\ 2s)^2\ (\sigma^*\ 2s)^2\ (\sigma\ 2s)^2\ (\pi_x\ 2p)^2\ (\pi_y\ 2p)^2\ ((\pi_x^*\ 2p)^2$ $(\pi_y^*\ 2p)^1$].

The $(\sigma\ 1s)^2$ and $(\sigma^*\ 1s)^2$ and do not take part in the chemical bonding. The $(\sigma\ 2s)^2\ (\sigma\ 2p)^2$, $(\pi\ 2p)^2$ and $(\pi_y\ 2p)^2$ are bonding orbitals and they contribute 8 electrons. The $(\sigma^*\ 2s), (\pi_x^*\ 2p)^2$ and $(\pi_y^*\ 2p)^1$ are antibonding orbitals and they contribute 2 + 2 + 1 or 5 electrons. Thus, the number of covalent bonds will be 1/2 (8 – 5) or 1.5 *i.e.*, the bond order is 1.5.

11. O_2^{2-} Ion : This type of ion, also called the peroxide ion, is present in sodium peroxide, NaO_2. This ion contains 18 electrons which are arranged in the following way :

$[(\sigma\ 1s)^2\ (\sigma^*\ 1s)^2]\ [(\sigma\ 2s)^2\ (\sigma\ 2p)^2\ (\pi_x\ 2p)^2\ (\pi_y\ 2p)^2\ (\pi_x^*\ 2p)^2\ (\pi_y^*\ 2p)^2]$

The bonding and antibonding p orbitals completely cancel each other, leaving one sigma bond from the filled σ 2p orbital and a bond order of one.

12. Fluorine molecule : The atomic number of fluorine is nine. It has electronic configuration $1s^2\ 2s^2\ 2p^5$. When fluorine molecule is formed, there will be 18 electrons in the F_2 molecule and these will be distributed in various molecular orbitals as :

$[(\sigma\ 1s)^2 (\sigma^*\ 1s)^2]\ [(\sigma\ 2s)^2 (\sigma^*\ 2s)^2 (\sigma\ 2p)^2 (\pi_x\ 2p)^2 (\pi_y\ 2p)^2 (\pi_x^*\ 2p)^2$ $(\pi_y^*\ 2p)^2\ ]$ $(\sigma\ 1s)^2$ and $(\sigma^*\ 1s)^2$ do not take part in bonding because only electrons in the valency shell take part. $(\sigma\ 2s)^2\ (\sigma\ 2p)^2\ (\pi_x\ 2p)^2\ (\pi_y\ 2p)$ are bonding orbitals and contribute 8 electrons. $(\pi_x^*\ 2p)^2, (\pi_y^*\ 2p)^2$ and $(\sigma^*\ 2s)^2$ form antibonding orbitals and they contribute, therefore, 6 electrons. Thus, the number of covalent bons is 1/2 (8 – 6) = 1. Thus, the bond between two fluorine atoms is a single bond. As all the electrons in bonding as well as in anitbonding orbitals are completely paired, it means that fluorine molecule is diamagnetic.

Similarly structures of other halogens (Cl, Br and I) can be explained.

13. Neon molecule : Its structure is similar to helium. In the formation of Ne_2, if possible sixteen outer electrons have to be accommodated in molecular orbitals. The electronic configuration of hypothetical Na_2 molecule is

$$[(\sigma\ 1s)^2\ (\sigma^*\ 1s)^2]\ [(\sigma\ 2s)^2\ (\sigma^*\ 2s)^2\ (\sigma\ 2p)^2$$
$$(\pi_x\ 2p)^2\ (\pi_y\ 2p)^2 (\pi_x^*\ 2p)^2\ \ (\pi_y^*\ 2p)^2\ ]\ (\sigma^*\ 2p)$$

Bond order in this case is (10 – 10) = 0. As antibonding MOs slightly dominate over the bonding MOs, N_2 molecule would be incapable of existence.

Bond Strength : From the above examples, it is evident that bond strength decreases with the increase in the number of antibonding orbitals. As nitrogen molecule does not involve antibonding orbitals (only valence orbitals are to be considered) whereas oxygen molecule does, it means that nitrogen is more stable than oxygen molecule.

Similarly, the stability of other molecules can be compared on the basis of number of antibonding orbitals used in their formation.

Energy Level Diagram for Molecular Orbitals : It is known that Is atomic orbitals of two atoms from two MOs designated as s (1s) and $\overset{*}{\sigma}$ and a (1s). The 2*s* and 2*p* atomic orbitals (eight AOs on two atoms) result in the formation of the following eight MOs :

Bonding MOs	:	s (2s)	s ($2p_z$)	π ($2p_x$)	π ($2p_y$)
		*	*	*	*
Antibonding MOs	:	σ (2s)	σ ($2p_z$)	π ($2p_x$)	π ($2p_y$)

Note : *The d-orbitals are not included in the present discussion since the purpose is to consider the formation of simple diatomic molecules only.

The energy levels of different molecular orbitals are shown diagrammatically. In this diagram, the molecular orbitals have been shown in the centre and the atomic orbitals which combine have shown on the two extremes.

The atomic orbitals of same energy have been shown at the same level while the bonding MOs are shown lower and the antibonding MOs have been shown higher than the corresponding atomic orbitals.

This indicates that while the bonding MOs are more stable, the antibonding MOs are less stable than the AOs from which they are formed. It is evident from the figure that *the extent of stabilization for a bonding MO is always equal to the extent of destabilisation for the corresponding antibonding MO.*

However, from spectroscopic studies it has been found that the above sequence of energy levels of MOs is not correct for all molecules. It has been observed that in case of homonuclear diatomic molecules of second row elements of the periodic table upto N_2 (viz. Li_2 Be_2, B_2, C_2, N_2), s ($2p_z$) *MO has been found to be higher in energy than the p (2px) and p* ($2p_y$) *MOs.* The reversal in the sequence of energy levels of MOs is ascribed to the positively of *mixing of 2s and $2p_z$ atomic orbitals,* as discussed below.

Mixing of Orbitals

It has been known that s (2s) and $\overset{*}{\sigma}$(2s) MOs get formed by the combination of 2s atomic orbitals of two atoms while σ ($2p_z$) and $\overset{*}{\sigma}$ ($2p_z$) MOs get formed by the combination of $2p_z$ atomic orbitals of two atoms.

However, if the energy differences between 2s and 2p atomic orbitals are small, there is a possibility of the *mixing of these orbitals* (cf. hybridisation of atomic orbitals) because of which the σ (2s) and $\overset{*}{\sigma}$ (2s) MOs do not retain pure s character and, similarly, the σ ($2p_z$) and $\overset{*}{\sigma}$ ($2p_z$) MOs do not retain pure p character.

On the other hand, *all the four orbitals acquire s mixed sp character.* Because of this s-p mixing, the energies of all the four orbitals change in such a way that the MOs s (2s) and $\overset{*}{\sigma}$ (2s) become more stable and thus get lowered in energy whereas the MOs $\sigma(2p_z)$ and $\overset{*}{\sigma}$ become less stable and are thus raised in energy.

As π (p) orbitals are not involved in mixing, the energies of p $(2p_x)$ and p $(2p_y)$ and p $(2p_y)$ MOs remain unchanged.

The s (2pz) *MO is raised to such an extent that it becomes higher in energy than p (2px) and p (2py) MOs.* The new sequence of energy of MOs, therefore, becomes as follows :

$$\sigma(1s), \overset{*}{\sigma}(1s), \sigma(2s), \overset{*}{\sigma}(2s), \pi(2p_x) = \pi(2p_y), \sigma(2p_z), \overset{*}{\pi}(2p_y), \overset{*}{\sigma}(2p_z)$$

—— Energy increase ⟶

The energy levels of different molecular orbitals obtained because of mixing of 2s and $2p_z$ atomic orbitals. Now, a question arises : How is it that the mixing of 2s and 2p atomic orbitals take place only in case of second row elements upto nitrogen only and not in case of oxygen and fluorine. The answer lies in the differences in energies of 2s and 2p atomic orbitals of atoms of second row elements, as depicted in the following Table 9.

Table 9 : Energy differences of 2s and $2p_z$ AOs of Atoms of Second Row Elements

Atoms	*Li*	*Be*	*B*	*C*	*N*	*O*	*F*
Difference in energies of 2s and $2p_z$ AOs (kJ $mole^{-1}$)	178	262	449	510	570	1430	1970

It is evident from the above table that while the differences in energies of 2s and $2p_z$ orbitals for atoms up to nitrogen are small, these differences becomes significantly large for oxygen and fluorine. Consequently, while the mixing of 2s and $2p_z$ AOs becomes readily possible in case of elements upto nitrogen, such a mixing in case of oxygen and fluorine has been practically ruled out.

It can be easily visualised that die to large energy differences, mixing of 1s and 2s atomic orbitals will also not take place. It is to be noted that out of the three 2p orbitals, only the $2p_z$ AO can combine with 2s atomic orbital due to symmetry considerations.

Filling up of Molecular Orbitals

The principles involved in filling up of electrons in molecular orbitals may sum up as below.

(i) The molecular orbitals with the lowest energy is filled first. Thus, s (1s) is filled first whereas s* (2p) is filled in the last. The order of filling up of molecular orbitals is as follows :

$$\sigma\,(1s) < \sigma^*\,(1s) < \sigma\,(2s)\ \sigma^*\,(2s) < \sigma\,(2p)$$
$$< \begin{pmatrix} \pi_x(2p) \\ \pi_y(2p) \end{pmatrix} < \begin{pmatrix} \pi_x^*(2p) \\ \pi_y^*(2p) \end{pmatrix} < \sigma^*\,(2p)$$

(ii) The maximum number of electrons in a molecular orbitals cannot exceed two,

(iii) The molecular orbitals are first singly filled up and paring starts only when more electrons are to be accommodated (*Hund's Rule*). Non-bonding orbitals are filled up just like bonding orbitals.

(iv) If there are two molecular orbitals at the same energy level, electrons will occupy separate orbitals in order to avoid each other, thus, leading to unpaired electron spins.

(v) *Bond order :* As an electron in an antibonding molecular orbitals cancels out the stability introduced by the electron in a bonding molecular orbital, it means that in order for bonding of atoms to occur there should be an excess of bonding electrons over antibonding electrons. In case where *the number of bonding and antibonding electrons are equal, no bond will be formed between the atoms.* However, the number of covalent bonds are related to bonding and antibonding electrons by a term called bond order which is defined as :

$$\text{Bond order} = \frac{\begin{array}{c}\text{No. of electrons in No. of electrons in} \\ \text{bonding orbitals} - \text{antibonding orbitals}\end{array}}{2}$$

Electronic Configurations of Heteronuclear Diatomic Molecules

Let us now consider the cases of heteronuclear diatomic molecules, in which the two atoms constituting a molecule are different. The principles involved in the distribution of electrons remain the same as discussed before.

However the molecular orbital diagram will not be symmetrical as in case of homonuclear diatomic molecules. Because of the difference in the electronegativites of the atoms, the electrons in bonding molecular orbital tend to spend more time near the more electronegative atom. On the other hand, the electrons in antibonding molecular orbital remain closer to the less electronegative atom.

Let us now discuss the electronic configurations and molecular orbital diagrams for some common hetero–nuclear diatomic molecules.

1. Carbon monoxide molecule, CO : The electronic configurations of C and O atoms are $1s^2\ 2s^2p^2$ and $1s^2\ 2s^2p^4$ respectively. Thus, there are 4 electrons in the outer shell of carbon and 6 electrons in the outer shell of oxygen. Hence, a total of 10 outer electrons have to be accommodated in the molecular orbitals of CO molecule.

Because of higher electronegativity of oxygen, its atomic orbitals would be of lower energy than the corresponding atomic orbitals of carbon. Because of this energy difference, the bonding and antibonding molecular orbitals will get more contribution from atomic orbitals of lower energy, *i.e.*, the atomic orbitals of oxygen and would be closer to it in energy than the antibonding molecular orbitals which would be closer to carbon in energy.

The bonding molecular orbitals would be having more characteristics of atomic orbitals of oxygen and antibonding molecular orbitals would have more characteristics of carbon.

The electronic configuration of CO molecule is.

$$[(\sigma\ 1s)^2\ (\sigma^*\ 1s)^2]\ \sigma\ (2s)^2\ \sigma^*\ (2\ s)^2\ \pi\ (2p_x)^2\ \pi\ (p_y)^2\ \sigma\ (2p_z)^2.$$

Bond order in this case is 1/2 (8 – 2) = 3. Thus, CO contains a triple bond *i.e.*, one sigma and two p bonds. Its bond dissociation energy has been found to be 1067 kJ mol^{-1} and bond length equal to 1.14Å..

From the molecular orbital energy level diagram, the following facts may be noted :

(i) We know that the electronegativity of oxygen is greater than that of carbon. Therefore, the 2s and 2p orbitals have been placed lower than the 2s and 2p orbitals of carbon.

(iii) We also know that CO is isoelectronic with N_2. Therefore, it has the same molecular orbital configuration as N_2 molecule has.

Hydrogen fluoride molecule, HF : The electronic configuration of hydrogen and fluorine have been $1s^1$ and $1s^2\ 2s^2\ p^5$, respectively. In the formation of HF molecule, only the 2p–electrons of fluorine atom would combine effectively with the solitary electron of hydrogen atom. It is known that only a p_z orbital is able to combine with an s-orbital.

The combination of a p_x *or* p_y orbital with an s-orbital is ruled out on symmetry considerations. The $2p_z$ orbital of fluorine yields a highly effective overlap with the s-orbital of hydrogen to form the σ (sp) molecular orbital.

It is known that the σ (*sp*) *is the only MO formed in HF molecule.* This is, evidently, a bonding orbital which is fully filled with 2 electrons. The rest of the electrons remain in their atomic orbitals. The MO formed in the case of HF molecule will not be symmetrical. Its shape. The asymmetry occurs because the energies of H (1s) and F ($2p_z$) atomic orbitals are not the same.

The MO diagram for HF molecule may be represented as in Fig. 5.64.

It can be seen that the bonding MO has an energy which lies closer to that of $2p_z$ orbitals of fluorine than to that of higher energy 1s–orbital of hydrogen.

4. Hydrogen Chloride molecule HCl : The electronic configurations of H and Cl atoms have been

$1s^1$ and $1s^2\ 2s^2p^6\ 3s^2p^5$, respectively. In this case also, there is only one bonding MO and one anti-bonding MO formed by the combination of 1s–orbital of hydrogen with $3p_z$ orbital of chlorine. The rest of the electrons of the chlorine atom remain in their respective AOs.

The electronic configuration of *HCl* is thus represented as follows :

$$\text{HCl : KL } (3s)^2\ \sigma(sp)^2\ (3p_x)^2\ (3p_y)^2$$

In this case also, the MO formed would not be symmetrical. The MO diagram of HCl molecule will be exactly similar..

Similarities Between Valence Bond and Molecular Orbital Methods

(i) When a bond is formed, both methods envisage an increase in charge density between the nuclei.

(ii) In both methods it is assumed that bond is only formed if orbitals of bond atoms overlap and have appropriate symmetry about the molecular axis.

(iii) In both methods, Aufbau's principle (Fault's principle and Hand's rule) is followed in filling atomic as well as molecular orbitals.

(iv) Both theories account for directional nature of covalent bonds.

Table 10 : Differences between molecular orbital and valence bond methods

Molecular Orbital Method	*Valence Bond Method*
1. In molecular orbital theory a molecule is composed of such atoms which lose their individual character.	1. In valence bond method, a molecule is composed of such atoms which do not lose their individual character when linked to other atoms.
2. An electron moves in the field of more than one nucleus, *i.e.*, the molecular orbitals are polycentric.	2. An electron moves in the field of one nucleus only, *i.e.*, atomic orbitals are monocentric.
3. When dealing with energy levels of electrons in molecules, electrons can be dealt with individually. are	3. When dealing with energy levels of electrons in molecules in the valence bond method, electrons dealt in pairs.
4. Molecular orbital theory accepts the ionic structures like $A^+ B^-$ and $A^- B^+$ in a molecule like AB.	4. Valence bond method rejects the ionic structures like $A^+ B^-$ and $A^- B^+$ in a molecule like AB.
5. Molecular orbital theory is more convenient for description of exited state in molecules.	5. It does not offer simple methods.
6. This theory readily explains the paramagnetism of the oxygen and sulphur molecules.	6. It does not readily explain.

From the above examples, it follows that there occurs a change in bond energy and bond length in the diatomic molecules. The data are listed in Table 11.

Table 11 : Characteristics of diatomic molecules.

Molecule	Li_2	B_2	C_2	N_2	O_2	F_2
Number of bonding electron	2	2	4	6	4	2
Bonding energy (kJ/mole)	105	290	630	945	496	151
Bonding length, Å	2.67	1.59	1.31	1.10	1.21	1.42

United Atom Method

In the LCAO method, atoms are brought from infinite distance gradually near one another till they take up a position of equilibrium in one molecule. Thus, the individual atomic orbitals yield molecular orbitals which may be bonding if the spins are paired or antibonding if the spins are unpaired.

In the united atom method, we start with an united atom in which the nuclei are superimposed over one another. Then, the individual nuclei are gradually separated till they acquire a position of equilibrium. It is important to remember that the united atom is purely hypothetical and has no actual existence.

Let us illustrate the formation of hydrogen molecule by the united atom method. In this treatment, we consider the structure of H_2 where the hydrogen nuclei are superimposed over each other.

In this structure there are no inter–atomic forces of attraction and repulsion. But there are naturally a great deal of repulsion between the two nuclei and electrons. Thus, the system will have higher energy and it will not be stable. If we now separate the nuclei very slowly, the repulsive forces will decrease and the energy of the system will be lowered. At equilibrium position, the two nuclei will be at a distance at which the forces of repulsion are minium.

This is also the position at which the energy of the system is lowest. This is the position at which the molecular orbital is formed. If we remove the atoms to infinity, the energy of individual atomic orbitals increase. Thus, we can represent the relationship between the orbitals in the united atom, molecular orbitals when the atoms are at equilibrium distance, and the individual atomic orbitals when atoms are at infinity as follows :

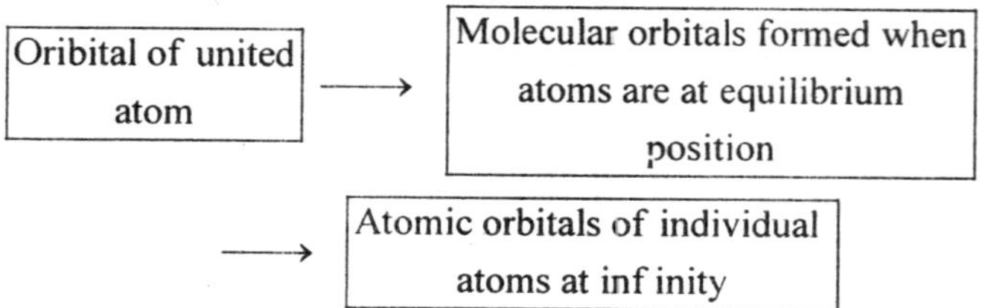

From the above, it is evident that molecular orbitals are intermediate in position between orbitals of united atom and orbitals of the separated atom. In there is a Mulliken correlation diagram for atoms which from a diatomic molecule.

Coulson's Mot Treatment of Co Molecule

According to coulson, the molecular diagram of CO given above is not correct and should involve the concept of hybridisation.

The electronic configuration of carbon a toms is $2s^2 2p_x^0\ 2p_y^1\ 2p_z^1$. The 2s and 2px orbitals undergo hybridisation to form two sp–hybrid orbitals which are represented as (sp). Thus, after hybridisation, the electronic configuration of carbon becomes as follows :

$$(sp)_c^2\ (sp)_c^1\ 2p_x^0\ \ 2p_y^0$$

Also 2s and $2p_z$ atomic orbitals on oxygen atom $(2s^2, 2p_x^2\ 2y_y^1\ 2p_z^1)$ undergo hybridisation to form two sp hybrid orbitals and the electronic configuration of oxygen atom then becomes as follows :

$$(sp)_0^2\ (sp)_0^1\ 2p_x^2\ \ 2p_y^2$$

As electron pairs $(sp)_0^1$ and $(sp)_0$ hybrid orbitals do not participate in bonding, these are known as σ–non bonding electron pairs (σ_{ab}– pairs). Thus, the CO molecule is having two lone pairs one pair on $(sp)_c$ and $(sp)_0$.

As $(sp)_c$ hybrid orbital having a lone pair has higher energy than that having a single electron, it means that it has a higher degree of p–character and is very reactive. This explains the properties of CO. Also $(sp)_c$ points away from O–atom.

As $(sp)_0$ hybrid orbital having lone pair has lower having lone pair has lower energy than that having a single electron, it has a high degree of s–orbital character and hence is very stable *i.e.* unreactive, hybrid orbital.

Now there remain two singly filed $(sp)_c$ and $(sp)_0$ hybrid orbitals and two 2p–orbitals ($2p_x$ and $2p_y$) on both the atoms which overlap to form the following orbitals.

(i) n $(s\ p)_c$ and $(sp)_0$ overlap to form $\sigma_{sp}{}^{+}$ and $\pi_{sp}{}^{*}$ orbitals.

(ii) $2p_x$ atomic orbital on carbon atom and $2p_x$ atomic orbital on oxygen atom overlap to form $p_x{}^{h}$ and $p_x{}^{*}$ molecular orbitals.

(iii) $2p_y$ atomic orbital on carbon atom and $2p_y$ atomic orbital on oxygen atom overlap to form $\pi_y{}^{b}$ and $\pi_y{}^{*}$ molecular orbitals.

The relative order of energy of these molecular orbitals and sp hybrid orbitals having lone pairs on each atom is as follows.

$$(sp)_0 < \sigma_{sp}{}^{b} < \pi_x{}^{b} = \pi_y{}^{b} < (sp)_c < \pi_x{}^{*}$$
$$= \pi_y{}^{*} < \sigma_{s\ p}{}^{*}$$

Therefore, the molecular orbital configuration of CO is as follows :

$$(sp)0^2,\ (s_{sp}{}^{b})^2,\ (p_x{}^{b})^2 = p_y{}^{b})^2,\ (sp)_c{}^2,\ (p_x{}^{*})^0,$$
$$= (\pi_y{}^{*})^0\ (\sigma_{sp}{}^{*})^0$$

The bond order of CO is 1/2 (6 – 0) or 3. This shows that carbon and oxygen in CO molecule are linked together by a triple bond (1s + 2p bonds).

Nitrosyl ion (N O) : The NO^+ ion has ten electrons in its valence shell. Therefore, it is isoelectronic with N_2 molecule which has also ten electrons. Thus, the NO^+ ion has the same molecular orbital configuration as N_2 molecule, *i.e.*,

$$(\sigma_{2s}{}^{b})\ (\sigma_{2s}{}^{*})\ (\pi_x{}^{b})^2 = (\pi_y{}^{b})^2,\ (\sigma_z{}^{b})^2.$$

Nitric oxide molecule, NO : The electronic configuration of N and O atoms have been $1s^2\ 2s^2\ p^3$ and $1s^2\ 2s^2\ p^4$, respectively. There are 5 electrons in the outer shell of nitrogen and 6 electrons in the outer shell of oxygen. Thus, a total of 11 outer electrons have to be accommodated in the molecular orbitals of NO molecule.

The electronic configuration of NO molecule is, thus as follows :

$$(\sigma\ ls)^2\ (\sigma^*\ 1s)^2\ \sigma\ (2\ s)^2\ \sigma^*\ (2\ \sigma)^2\ \sigma\ (2p_z)^2\ \pi\ (2p_x)^2\ \pi\ (2p_y)^2\ \pi^*(2p_x)^1.$$

The bond order in this case is 1/2 (8 – 3) = $2\frac{1}{2}$. This represents one σ–bond and two π–bonds less the antibonding effect of one unpaired electron in the π^* $(2p_x)$ molecular orbital, the presence of this antibonding electron makes the NO molecule less stable than the N_2 molecule.

This has been confirmed by the fact that the bond dissociation energy of nitric oxide, viz. .667.8 kJ mol^{-1} has been less than that of nitrogen which is 945.6 kJ mol^{-1}.

As NO contains an unpaired electron, it is *paramagnetic.*

As in the case of CO molecule, the bonding molecular orbital would be having greater characteristics of atomic orbitals of oxygen and the antibonding molecular orbitals would have greater characteristics of atomic orbitals of nitrogen.

The bond order 2½ of nitric oxide indicates that this molecule is paramagnetic.

2

Schematic Representation of the Ionic Bonds

INTRODUCTION

The most striking example of electrovalent bonds is the combination of sodium and chlorine to form sodium chloride. Sodium (2, 8, 1) possesses one surplus electron by losing which it is changed into a sodium ion getting the stable configuration of the inert gas neon (2, 8). On the other hand, chlorine (2, 8, 7) needs one electron by gaining which it is changed into chloride ion acquiring the stable configuration of the inert gas argon (2, 8, 8). The formation of sodium chloride may be depicted below.

$$\underset{2,\,8,\,1}{Na^{\times}} + \underset{2,\,8,\,7}{\cdot\ddot{\underset{..}{Cl}}:} \longrightarrow \underset{2,\,8}{Na}\ \underset{2,\,8,\,8}{\left[{}^{\times}_{\cdot}\ddot{\underset{..}{Cl}}:\right]^{-}}$$

One can explain the formation of magnesium chloride in a similar manner as in sodium chloride. The electronic configuration of magnesium is 2, 8, 2 and that of chlorine is 2, 8, 7. The magnesium atom loses two electrons, one each to two chlorine atoms. The resulting magnesium ion, therefore, carries two units of positive charge while each chlorine atom acquires one negative charge is shown. Obviously, the electrovalency of magnesium is +2 while that of chlorine is –1.

Similarly, one can explain the combination of metals like Li, K, Rb, Cs, Ca Ba, etc., with F, Cl, Br, I,S,O, etc. *the number of electrons lost or gained by an element during the formation of an electrovalent bond is termed as valency number.*

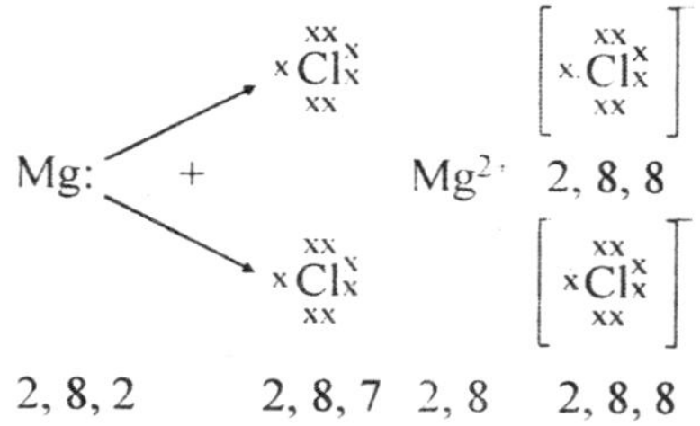

$$\mathrm{Mg:} \begin{matrix} \nearrow \overset{xx}{\underset{xx}{{}_{x}Cl_{x}^{x}}} \\ + \\ \searrow \overset{xx}{\underset{xx}{{}_{x}Cl_{x}^{x}}} \end{matrix} \qquad Mg^{2+} \begin{matrix} \left[\overset{xx}{\underset{xx}{{}_{x}Cl_{x}^{x}}}\right]^{-} \\ 2, 8, 8 \\ \left[\overset{xx}{\underset{xx}{{}_{x}Cl_{x}^{x}}}\right]^{-} \end{matrix}$$

2, 8, 2 2, 8, 7 2, 8 2, 8, 8

The valency number may have positive or negative sign. For example, Na, Ca and Al have + 1, + 2 and + 3 whereas Cl, S and N have –1, –2 and –3 as their valency numbers respectively.

FACTORS GOVERNING THE FORMATION OF AN IONIC BOND

The formation of an ionic bond is governed by the ease with which anions and cations are formed. This further depends upon.

(a) Ionisation energy : One of the two atoms forming an electrovalent bond should have a *low* level. It means that the lower the vale of ionisation potential, the greater will be the ease of formation of cations and hence of electrovalent bond.

$$M \xrightarrow{E_1} M^{+} + e^{-} \; [E_1 = \text{First ionisation energy}]$$

$$\xrightarrow{E_2} M^{2+} + e^{-} \; [E_2 = \text{Second ionisation energy}].$$

The alkali metals (group I) have low ionisation energies. Therefore, they can readily lose an electron from their valency shells. On the other hand, the alkaline earth metals, do so less readily.

The reason is that atoms of alkaline earth metals have to lose two electrons from their outer energy level so as the acquire stable configuration of nearest ineart gas.

Once an electron has been lost from an atom, the remaining electrons are held more firmly and are not lost so easily. Thus, second ionisation energy (E_2) is very high as compared with the first ionisation energy (E_1). Therefore to remove two electrons, energy equal to $E_1 + E_2$ is required. Because of this reason alkaline earth metals with two electrons in the outermost energy level form positive ions less readily.

The removal of three or more electrons needs large energies. Therefore positive ions with three or four units are rarely formed.

(b) Electron affinity : One of the two atoms forming an electrovalent bond should have a *high electron affinity, i.e.*, it should be able to accept an electron in its outer energy level readily. It means that the higher the electron affinity, the greater the ease of formation of the anion. The elements of group VII (halogens) posses the highest electron affinities of all the elements and hence they are converted into anions by accepting electrons easily and form ionic bonds readily.

The elements of group VI have quite high electron affinities and form ionic bonds but less readily than those of VII group elements because they accept two electrons to achieve stable configuration of nearest inert gas. The reason for this is that the addition of one electron to the atom is accompanied they release of energy to form a singly charged negative ion. The negative ion so formed repels the new coming electron and energy has to be supplied to add electron to overcome the strong repulsion from the already negatively charge ion.

$$\underset{\text{Atom}}{X + e^-} \longrightarrow \underset{\text{Singly charged ion}}{X^- + \text{Energy}} \quad \text{(Electron affinity)}$$

$$\underset{\text{Singly charged ion}}{X^-} + e^- + \text{energy supplied} \longrightarrow X^{2-}$$

So the formation of doubly–charge negative ion does not take place with ease. Therefore, formation of oxides or sulphides is not so easy as the formation of fluorides or chlorides. Thus, it may be concluded that the ionic bonds are generally formed between the atoms of the elements of the left of the periodic table.

(c) Lattice energy : The formation of an ionic bond depends upon the lattice energy. The higher the value of lattice energy of resulting ionic compound, the greater will be its ease of formation. For lattice energies to be high, the force of electrostatic attraction between the constituent ions should be high.

Electrovalent Compounds : Compounds involving electrovalent bonds are called *electrovalent compounds.* These are also *called polar compounds* since on dissolution in water they split up into two types of oppositely charge poles called ions.

CHARACTERISTICS OF ELECTROVALENT COMPOUNDS

(a) **Crystalline nature :** Ionic compounds are usually crystalline in nature. X–ray diffraction patterns have proved that the constituent units in a crystal are ions and not molecules. These ions are arranged together in a regular way in a lattice. The attraction between ions is *electrostatic and non-directional.*

For example, in a crystal of sodium chloride, every sodium ion is surrounded by six evenly-spaced chloride ions and every chloride ion by six evenly spaced sodium ions in a regular fashion. This arrangement extends throughout the crystal lattice resulting in strong electrostatic forces throughout the crystal

(b) **Melting and boiling points :** Electrovalent compounds possess high melting and boiling points. Their relative high melting and boiling points are due to electrostatic forces of attraction and, considerable amount of energy is needed to break up their crystal lattices.

The closer the ions in a crystal, the stronger will be the electrostatic attraction forces and consequently higher will be the melting points (as shown in Table 1).

Table 1 : Melting points of some sodium salts.

Anion	F^-	Cl^-	Br^-	I^-
Melting point (°C)	995	800	750	691

(c) **Conductance of compounds :** *Ionic solids have low conductance of electricity in solid state.* The reason is that the anions and cations on account of electrostatic forces remain intact occupying fixed positions in the crystal lattice. The ions, therefore, are unable to move to any large extent when an electric field is applied. Hence little current flows. However, a few solids like M Ag_5 I_5 (M = K, NH_4, Rb) are known that have conductivity similar to those of the ions in solutions due to the high mobility of Ag^+ ions in an unusual crystal lattice.

When electrovalent compounds are melted or dissolved in a solvent of high dielectric constant the binding forces in the crystal lattice disappear and the component ions become *mobile.* As the ions are charged, they are attracted towards the electrodes and thus act as carrier of the electric current. It means that the melts or solutions of electrovalent compounds conduct electricity.

(d) **Stability :** Electrovalent compounds are quite soluble in solvents of high dielectric constant and of polar nature. It is because the high dielectric constant of the solvent lowers the attractive forces between the ions of the lattice. Also, because of the polar nature, the solvent interacts highly with the ions and the energy of solution is sufficient to overcome the attractive forces between the ions. Non-polar solvents like benzene, carbon tetrachloride, etc., which have low dielectric constants, do not solvate the ions. Hence ionic compounds are soluble in polar solvents but insoluble in non-polar solvents.

(e) **Ionic reactions :** Chemical reactions of an ionic compound are the reactions of its constituent ions. These are usually rapid.

$$\underset{\text{Acid}}{H^+} + \underset{\text{Base}}{OH^-} \rightarrow H_2O \text{ (Neutralisations reaction)}$$

$$Ag^+ + Cl^- \rightarrow AgCl \text{ (Precipitation reaction)}$$

(f) **Space–isomerism :** The linkage between the two oppositely charge atoms is simply due to the electrostatic attraction between them and is, therefore, *non-rigid and non-directional.* They are incapable of exhibiting any type of space–isomerism.

(g) **Isomorphism :** Mitscherlich's law could not explain the isomorphism of compounds like sodium fluoride and magnesium oxide or potassium sulphide and calcium chloride. However, the electronic theory of valency offers explanation of suggesting that compounds showing isomorphism have similar electronic structures as shown below.

(i) **Sodium fluoride** **and** **Magnesium oxide**

Na^+	F^-		Mg^{2+}	O^{2-}
2, 8	2, 8		2, 8	2, 8

(ii) **Postassium sulphide** **and** **Calcium chloride**

K^+	S^{2-}	K^+	Cl^-	Ca^{2+}	Cl^-
2, 8, 8	2, 8, 8	2, 8, 8	2, 8, 8	2, 8, 8	2, 8, 8

(h) **Brittleness :** If a sufficient amount of energy is taken up by the crystal layer of unit cell, then anion-anion and cation-cation repulsive forces become greater than attractive forces and thereby the crystal crumbles. Thus, ionic solids, though hard, are brittle and can be powdered easily by hammering.

Variable Electrovalency : There are a number of elements (both normal and transition) which show more than one valency in the formation of their compounds. There are a number of reasons for this variable valency. Some of which are given below.

1. Outer shell configuration of ions is not stable : It may be noted that alkali and alkaline earth metal ions have most stable s^2p^6 configuration in their ionic compounds (Na^+, Ca^{2+} etc.). Lithium ion has. s^2 (helium configuration) and this is also highly stable. The configuration of these cations are shown below:

Ions	*Configuration*	*Ions*	*Configuration*
Li^+	Is^2		
Na^+	$1s^2\ 2s^2\ 2p^6$	Mg^{2+}	$1s^2\ 2s^2\ 2p^6$
K^+	$1s^2\ 2s^2\ 2p^6\ 3s^2\ 3p^6$	Ca^{2+}	$1s^2\ 2s^2\ 2p_6\ 3s^2\ 3p^6$

Such stable structure should hardly lose any more electrons to form another ion with a different valency. Thus, the normal *elements do not exhibit variable electrovalency.*

Ions of transition metals do not have the stable $s^2\ p^6$ configuration in their outermost shells. The number of outer most electrons is more than 8 electrons and some time goes upto 18 (*e.g.*, copper group). *In these cases outermost shell is not stable and hence ions are also not stable. The outer shell loses one or more electrons, giving rise to metal ions of higher valencies.* A few examples are given below.

(a) ns^2p^6 *dx* type configuration where the sum (2 + 6 + x) should be more than 8 and less than 18. The cations possessing this type of configuration are transitional metal cations *e.g.*, Fe^{2+}, Co^{2+} and Ni^{2+}

Metal	*Configuration of atoms*	*Ion*	*Configuration of ions*
Fe	2, 8 $3s^2\ 3p^6\ 3d^6\ 4s^2$ 14 electrons	Fe^{2+}	2, 8, $3s^2\ 3p^6\ 3d^6$
Co	2, 8 $3s^2\ 3p^6\ 3d^7\ 4s^2$ 15 electrons	Co^{2+}	2, 8, $3s^2\ 3p^6\ 3d^7$
Ni	2, 8, $3s^2\ 3p^6\ 3d^8\ 4s^2$ 16 electrons	Ni^{2+}	2, 8, $3s^2\ 3p^6\ 3d^8$

The above outer configuration $3s^2\ 3p^6\ 3d^x$ where x = 6, 7, 8 is present in Fe^{2+}, Co^{2+} and Ni^{2+} ions respectively. This is not stable as the s^2p^6 configuration. Hence, these ions can easily lose one of the *d* electrons, producing trivalent ions, *viz.*, Fe^{3+}, Co^{3+} and Ni^{3+} ions.

The trivalent ions are more stable since the *effective positive charge* on the nucleus is now higher and prevents further removal of any more electrons. Hence atoms of these elements have a tendency to lose different number of electrons under different conditions.

Thus, these elements show two or more different valencies, *i.e.*, they show variable valency. But these elements only form trivalent ions, Fe^{3+}, Co^{3+} and Ni^{3+}.

Ions with more than three units of charge are not formed. It is because now the higher effective positive charge on the nucleus checks the tendency of the positive trivalent ions to lose more electrons due to increasing ionisation energy.

(b) *Pseudo inert gas configuration* ($ns^2p^6d^{10}$) *i.e., cations with 18 electrons in their outer shell.* This type of configuration is found in transition metal ions.

Consider, copper silver and gold,

Metal	*Configuration of atoms*	*Ion*	*Configuration of ions*
Cu	2,8, 18, $3s^2$, $3p^6$ $3d^{10}$ $4s^1$ 18 electrons	Cu^+	2, 8, $3s^2$ $3p^6$ $3d^{10}$
Ag	2, 8, 18,$4s^2$ $4p^6$ $4d^{10}$ 18 electrons	Ag^+	2,8, 18, $4s^2$ $4p^6$ $4d^{10}$
Au	2, 8, 18, 32, $5s^2$ $5p^6$ $5d^{10}$ $9s^1$ 18 electrons	Au^+	2, 8, 18, 32, $5s^2$ $5p^6$ $5d^{10}$

Here $ns^2p^6d^{10}$ configuration in Cu^+, Ag^+ and Au^+ is less stable than the $ns^2\ p^6$ configuration and hence can lose one or more electrons to yield cations of higher valency.

Thus, Cu^+ can give Cu^{2+}, Ag^+ can give Ag^{2+} while Au^+ loses two electrons to yield Au^{3+}; Cu^{2+} and Au^{3+} ions are quite stable while Ag^{2+} is not very stable.

The divalent Zn^{2+}, Cd^{2+}, and Hg^{2+} ions also have 18 electrons in their outer shell like Cu^+, Ag^+ and Au^+ ions but do not give cations of higher valency states.

Cu^+ and Zn^{2+} (18 electrons),

Ag^+ and Cd^{2+} (18 electrons),

Au^+ and Hg^{2+} (18 electrons).

But the nuclear charge in Zn^2 ions. Hence in the former ions all the 18 electrons are held firmly by nucleus so there is no further loss of electrons. Consequently, Zn, Cd and Hg cannot form ions with valency more than two.

2. Inert pair effect : Some of the heavier elements which are having three or four electrons in their outermost shell are expected to form trivalent or tetravalent ions. But some times they do not lose electrons and hence form cations with lower positive charge. Such electrons which do not participate in ion formation are called inert parts. This effect is called inert part effect. When inert pair is also lost under suitable conditions, higher charged cations are formed. A few examples are given below Table 2.

It is clear from above discussion that in the formation of Sn^{2+} and In^+ ions, the pair of 6s-electrons and in the Tl^+ ion the pair of **6s**-electrons do not take part in bond formation. The reluctance of the s-electron pair to take part pin bond formation is known as *inert pair effect.*

Table 2

Atom or ion	*Configuration*	*Remarks*
Sn	2, 8, 18, 18, $5s^2$ $5p^2$	There are four valency electrons.
Sn^{2+}	2, 8, 18, 18, $5s^2$	Only two *p* electrons are lost. The 5*s* electron pair does not involve in bond formation.
Sn^{4+}	2, 8, 18, 18	All the four electrons are involved in bond formation.
Tl	2, 8, 18, 32, 18, $6s^2$, $6p^1$	There are three valency electrons.
Tl^+	2, 8, 18, 32, 18, $6s^2$	Only one 6p-electron is lost; 6s-electron pair is not involved in bond formation.
Tl^{3+}	2. 8, 18, 32, 18	All the three valency electrons are involved in bond formation.
In	2, 8, 18, 18 $5s^2$, $5p^1$	There are three valency electrons.
In^+	2, 8, 18, 18, $5s^2$	Only one *5p* electron is involved in bond formation.
In^{3+}	2, 8, 18, 18	All the three electrons $5s^2$ $5p^1$ are involved in the formation of bond.

The inert pair effect is more pronounced in case of heavier elements. As a result of this there is less tendency of the formation of Sn^{4+}, Tl^{3+} and In^{3+} ions.

Inert pair effect is also shown by P, As, Sb and Bi, the elements of group V.

P : .... $1s^2\ 2s^2\ 2p^6\ 3s^2\ 3p^3$

As : .... $3s^2\ 3p^3\ 3d^{10}\ 4s^2\ 4p^3$

Sb : .... $4s^2\ 4p^6\ 4d^{10}\ 5s^2\ 5p^3$

Bi : .... $5s^2\ 5p^6\ 5d^{10}\ 6s^2\ 6p^3$

All these elements show valency of three (when s-electron pair remains inert) as well as five (when all the five electrons are involved). The inert pair effect is most marked in bismuth which is heaviest element of this group. Thus, bismuth only forms $BiCl_3$ while other elements form trivalent as well as pentavalent chlorides.

A partial explanation of the inert pair effect is due to the fact that the outer ns^2-electrons penetrate to some extent the 18 electron shell immediately after it. ns^2 electrons are, therefore, less effectively screened from the attractive power of the nucleus and so are bound more firmly to the atom.

SOLVATION ENERGY

Dissolution of Electrolytes (Ionic Compounds)

It is found that different electrolytes tend to dissolve to different extent in a solvent (say, H_2O at the same temperature). It could be explained with the help of following factors.

(i) Dielectric constant of solvent : According to Coulomb's law of electrostatic forces of attraction (F) we have

$$F = \frac{q_1 q_2}{r^2 D}$$

where, D = Dielectric constant of Solvent

For H_2O, D = 80

For C_2H_5OH, D = 25

As the dielectric constant of water has been very high as compared to that of (C_2H_5OH), the forces of attraction between Na^+ and Cl^- ions in the crystal get reduced to minimum. Thus, NaCl is having more solubility water than in ethyl alcohol (C_2H_5OH). Its N^+ and Cl^- ions are hydrated and a lot of hydration energy gets released. The system thus, tends to attain minimum energy and maximum stability.

(ii) Lattice energy : The cations and unions in an ionic solid are held together by strong ionic attractive forces. So as to dissolve the ionic solids, these forces must be cut off. For this purpose, energy has to be supplied to separate the ions.

The energy required to separate the ions in one mole of the ionic solid is termed as its *lattice energy. Smaller the lattice energy, greater the tendency of an ionic solid to dissolve.* Hence,

$$\text{Lattice energy} \propto \frac{1}{\text{Tendency of an ionic solid dissolve}}$$

$$\underset{\text{One mole}}{NaCl^{-}} + \underset{\text{766 50 k cals}}{\text{Lattice energy}} \rightarrow \underset{\text{One mole}}{Na^{+}(g)} + \underset{\text{One mole}}{Cl^{-}(g)}$$

We know that $PbSO_4$ is sparingly soluble while Na_2SO_4 is highly soluble in water. The reason is that, the lattice energy of $PbSO_4$ is very high as compared to Na_2SO_4

(iii) Solvation of ions : When we add an ionic solid to a solvent, the cations and anions present in this solid are respectively surrounded by the negative and positive ends of the solvent molecules.

Surrounding solvent molecules, partly tend to neutralise the charge of the ions and also act as an insulating sphere. The insulating effect is termed *as dielectric effect.*

Thus ion-dipole interaction which causes loose combination between the ion and solvent molecules is termed as *solvation.* If the solvent has been water, it is termed as *hydration.* During solvation (hydration), energy always gets released which is termed as *solvation energy (hydration energy).*

Thus, solvation (hydration) energy refers to the energy released when one mole of the ions are solvated (hydrated).

$$\underset{\text{One mole}}{K^{+}(g)} + \text{water} \rightarrow \underset{\text{One mole of hydrated ions}}{K^{+}(aq)} + \text{Hydration energy.}$$

Higher the hydration energy, higher the solubility of the ionic solid. *Ions with small size and high charge are heavily hydrated and are having maximum hydration energy, e.g.,* the hydration energy of Li^{+} ion has been more than that of Na^{+} ion which in turn has been greater than that of K^{+}–ion.

Overall Tendencies for Dissolution

From the above discussion, it can be concluded that lattice energy and hydration energy have been mainly responsible for dissolution of an ionic solid. The overall energy change due to these energies is termed as *heat of solution,* ΔH.

Thus, ΔH (solution) = Hydration energy – Lattice energy.

When hydration energy has been greater than lattice energy, the system will acquire minimum energy and maximum stability. Hence, the solubility of the ionic solid would tend to increase.

It is to be noted that the magnitude of lattice energy and hydration energy depends upon the same factors. However, with increase in charge and decrease in size of ions, increase in lattice energy has been more as compared to hydration energy. Thus, with increase in charge and decrease in size of the ions, solubility of electrolytes tend to decrease. For instance Na^+Cl^- has been more soluble in water than Mg^{2+} O^{2-}.

$AlCl_3$ has been covalent compound in the dry state while it has been completely ionic in the aqueous solution.

Reason : If aluminium chloride, $AlCl_3$ (dry) is treated with water, it is easily hydrated. Due to hydration , energy is released which is termed as *Hydration Energy.* This energy is sufficient for removing three electrons to form Al^{+3}, ion

$$AlCl_3 + 6H_2O \longrightarrow AlCl_3 6H_2O + \text{Hydration energy}$$
$$\downarrow\uparrow$$
$$[Al(H_2O)_6]^{+3} + 3Cl^-$$

LOW SOLUBILITIES OF SILVER HALIDES

The low solubility of compounds like lead sulphate barium, sulphate, etc., could be explained on the basis of their high lattice energy. However, the solubilities of AgCl, AgBr and AgI in water could not be explained on the same basis. The reason that the lattice energy of these compounds has been almost of the same order as that of highly soluble alkali metal halides. This unusual behaviour could be explained on the basis of the outer electronic configuration of the cations which are under study.

The outer electronic configuration of Ag^+ ion may be put as follows

$$(n - 1)\, d^{10}\, ns^0\, np^0. \text{ Here, } n = 5$$

The outer electron configuration of alkali metal ions like Na^+. K^+, etc., would be ns^2 np^6. We know that the d-electrons are able to screen the nucleus from outer electrons less effectively than *s* and *p* electrons. The Ag^+ ion is having *d* electrons whereas alkali metal ions do not have. Thus, the nucleus of Ag^+ ion will attract its outer electrons more strongly as compared to alkali metal ions. Because of this, silver halides undergo appreciable polarisation and hence appreciable change from ionic to covalent character. Thus silver halides have been less soluble in water.

HIGHLY CHARGED CATIONS ARE RARE

Highly charged ions, tripositive and tetrapositive, would be obtained by the removal of three and four electrons respectively from the neutral atom. In order to get these very high ionisation energy is required.

Such a high energy would not be available in most of the chemical reactions. Nearly 1470 kJ/mole energy is available for removing two electrons from neutral gaseous atom. Hence dipositive ions are generally formed.

Al^{+3}(g) ion in not formed. The reason is that 5132.4 kJ/mole energy is required to remove three electrons from Al-gaseous atom.

$$Al(g) + 579.6 \text{ kJ/mole} \rightarrow Al^+(g) + e^-$$

$$Al^+(g) + 2039.6 \text{ kJ/mole} \rightarrow Al^{+2}(g) + e^-$$

$$Al^{+2}(g) + 2709.0 \text{ kJ/mole} \rightarrow Al^{+3}(g) + e^-$$

$$Al(g) + 5132.4 \text{ kJ/mole} \rightarrow Al^{+3}(g) + 3e^-$$

As 5132.4 kJ/mole energy has been much higher than 1470 kJ/mole (available in ordinary chemical reactions), Al^{+3}(g) ion could not be formed.

HIGHLY CHARGED ANIONS ARE RARE

Highly charged ions such as tri-negative and tetra-negative could be obtained by the addition of three and four electrons respectively into the neutral atom. For getting these, very high energy has to be supplied. Such a huge amount of energy cannot be supplied as the reaction in energetically unfavourable. N^{-3} ion and C^{-4} ions could not be formed due to the above reason. It is to be noted that.

(i) Addition of first electron causes the release of energy.

(ii) Addition of second and third electron causes the absorption of energy. It occurs due to repulsion between the negative ion and the added electrons.

COVALENT BOND OR NON-POLAR BOND

The formation of ionic compounds has been explained on the basis of transfer of electrons and Coulomb's law of attraction between oppositely charged particles.

So as to understand the formation of polar and non-polar compound, we will make an attempt to study the concept of covalent bond based on sharing of electrons between atoms.

When both the atoms (similar or dissimilar) taking part in a chemical combination are short of electrons than the nearest inert gas configurations, the combination between them takes place by sharing of electrons.

Each atom contributes one electron to form a common pair which then is shared by both. *Bond established between atoms by this process of sharing is known as a covalent bond.*

Consider two atoms A and B which are short of one electron than the stable neon gas structure. If the two atoms share a pair of electrons between them, by dong so both of them acquire neon structure as shown below.

$$:\ddot{\underset{..}{A}}\cdot + {}_{x}\overset{xx}{\underset{xx}{B}}{}^{x}_{x} \longrightarrow :\ddot{\underset{..}{A}}\,{}^{\cdot}_{x}\overset{xx}{\underset{xx}{B}}{}^{x}_{x} \quad \text{or} \quad A-B$$

A covalent linkage is expressed in formula by a line (—). A covalent linkage is non-polar and non-ionised because in its formation electrons are not transferred from one atom to the other, and thus the molecule does not acquire polarity and no ions are formed.

Electronic Theory of Covalent Bond

G. N. Lewis (1916) introduced "Electronic Theory" to explain the concept of covalent bond between atoms. According to this theory,

(i) A covalent bond would be formed between two similar non-metallic atoms A-A or B-B or dissimilar atoms (A and B).

(ii) Both the atoms A and B have been short of one or more electrons than the nearest inert gas configuration.

(iii) Both the atoms A and B have been sharing one or more electrons. The shared pair becomes their common property. Because of this atoms attain the inert gas configuration.

For *homonuclear molecules* (A-A or B-B), the bond formed between them is termed as non-polar covalent bond.

For heterouclear Molecules, (A–A), the bond formed is termed as polar covalent.

(iv) If both the atoms have been sharing one, two or three electrons each, single, double or triple covalent bonds would be formed respectively

(v) The covalent bond has been a directional bond.

(vi) There always occurs a decrease in energy during the formation of a molecule. Thus the molecule becomes stable.

Single and Multiple Covalent Bonds

A bond B is said to be a *single covalent bond* if the atoms share only one pair of electrons. Similarly, covalent bonds formed by the sharing of two or three electron pairs, between atoms, are known as *double or triple bonds* respectively.

Covalency : It is defined *as the number of electrons that an atom contributes for sharing purpose in order to complete its extreme orbit, e.g.,* carbon atom contributes all its four valency electrons for sharing purposes to complete its octet. Hence covalency of carbon is four. Similarly, the covalency of H, 0 and N atoms is 1, 2 and 3 respectively.

Compounds formed through covalency are called **covalent** or **non-polar** *since they do not ionise in water.*

Following examples of electron dot representation of molecules are given for clarity.

(i) Formation of hydrogen molecule : In the formation of hydrogen molecule, each hydrogen atom provides one electron and then the two electrons are shared between the two atoms so that both the atoms acquire stable configuration. This results in the formation of a single covalent bond.

Thus, covalency of hydrogen is one.

(ii) Formation of chlorine molecule : Chlorine atom has seven electrons in the outermost orbit. In the formation of chlorine molecule, two atoms contribute one electron each for sharing and thus acquire a stable configuration of argon. Thus, the two chlorine atoms are held by a single covalent bond.

Thus the covalency of chlorine is one.

(iii) **Formation of oxygen molecule :** As oxygen atom has six electrons in its outer most orbit, it, means that two oxygen atoms contribute two electrons each for sharing so that a stable molecule is formed.

Thus covalency of oxygen is two.

This type of covalent bond which is formed by the mutual sharing of two pairs of electrons between two atoms is called double bond and is indicated by two lines (=). However, this structure is not correct.

(iv) **Formation of nitrogen molecule :** Nitrogen atom has five electrons in its outermost orbit; two atoms contribute three electrons each for sharing and resulting in the formation of a stable molecule.

Thus covalency of nitrogen is there.

This type of covalent bond which is formed by the mutual sharing of three pairs of electrons between two atoms is called a *triple bond* and is indicated by three lines (≡).

Some other examples involving single, double and triple bond are given below.

(iii) Formation of ethylene molecule.

$$\mathrm{H\dot{x}\ \overset{\overset{H}{x.}}{C}::\overset{\overset{H}{.x}}{C}x\ \ H\ or\ H-\overset{\overset{H}{|}}{C}=\overset{\overset{H}{|}}{C}-H}$$

(iv) Formation of acetylene molecule.

$$\mathrm{H^{x}_{.}\ C:::C^{x}_{.}\ H\ \ H-C\equiv C-H}$$

(v) Formation of hydrogen chloride.

$$\mathrm{H^{x}\ +\ :\ddot{\underset{..}{Cl}}:\rightarrow X^{\bullet}_{x}\ \ddot{\underset{..}{Cl}}:\ \ or\ \ H—Cl}$$

Characteristics of Covalent Compounds

(a) **Physical state :** Generally, covalent compounds are gases or liquids at room temperature. They are only solids when their molecular weights are very high. For example chlorine (mol. wt. = 74) is a gas, bromine (mol. wt. = 160) is a liquid and iodine (mol. wt. = 254) is a solid.

(b) **Crystal structure :** The crystals of covalent compounds are generally of three types:

(i) In this first type, there are crystals in which the unit is the molecule. The molecules are held by vander Waal's forces which are .much weaker than electrostatic forces found in ionic crystals. The covalent crystals of this type are readily fusible and volatile. Examples of such crystals are furnished by inorganic substances like sulphur, iodine, phosphorus pentoxide, etc.

(ii) The second type of crystals formed by covalent compounds include giant molecules in which each carbon atom is linked with four neighbouring carbon atoms held at the corners of a regular tetrahedron by covalent bonds. Since the covalent bonds C–C are very strong and run without a break through-out the whole crystals, resulting hardness in them. Examples of such crystals are those of diamond, silicon carbide, aluminium nitride, etc.

(iii) The third type includes crystals which have separate lattice layers. An interesting example is graphite. In diamond every atom is at the same distance from all its four neighbours whereas in graphite it has the same relation as in diamond to only three of them, but is more than double the distance from the fourth.

Hence, in graphite there are a series of layers, each-one atom thick, that are readily separable from one another. This makes graphite as an admirable lubricant.

(c) **Melting and boiling points :** For melting a covalent compound where van der Waal's forces are operative, only a small amount of thermal energy is needed to overcome these forces. Hence, such a covalent compound has low melting and boiling point.

For covalent solids like graphite and diamond where strong forces operate, a large amount of thermal energy is required to overcome such forces. Hence, such crystals have high melting and boiling points.

(d) Solubility : Covalent compounds, in which van der Waal's forces are operative, are normally insoluble in polar solvents of high dielectric constant but soluble in non-polar organic solvents of low dielectric constant, such as benzene and carbon tetrachloride, *i.e.*, like dissolves like. Covalent compounds having giant molecules are practically insoluble in all solvents due to the big size of the molecules constituting such compounds.

(e) Conductivity : Covalent compounds having giant molecules or simple molecule units do not conduct electricity in the fused as well as in the solution state as there are no free electrons or ions to carry the current. On the other hand, substances like graphite which consist of separate layers conduct electricity as the electrons can pass between the two layers.

(f) Reaction rates : Reaction rates between covalent molecules are much slow compared to those between ionic compounds because the former involves breaking of covalent bonds and establishing of new bonds while in the latter there is only re-grouping of ions.

(g) Isomerism : Except *s* orbitals, other orbitals (*p*, *d* and *f*) are oriented in space. As the covalent bonds are formed by the overlapping of atomic orbitals, the resultant orbitals of the bonded atoms must also be concentrated along certain directions. This imparts directional nature to covalent bonds and due to the directional nature, they show isomerism (structural and space).

EXPLANATION OF THE FAILURE OF OCTET RULE

It is possible to explain the failure of octet rule by the help of the following concepts.

(i) Sidgwick's rule of maximum covalency : In the octet theory it is assumed that the maximum number of electrons in the valency shell is 8. However, Sidgwick believes that this number can be exceeded. Actually, the maximum covalency of element actually depends on the period in which the element concerned is present. For example, in SF_6 sulphur atom will have a valency group of 12 electrons. Similarly in OsF_8, osmium will have a valency group of 16 electrons and in PCl_5 the phosphorus atom would have 10 electrons.

On the basis of Sidgwick's rule, the structure of PCl_5.

(ii) Singlet linkage : Contrary to the views of Sidgwick, *Sugden* suggested that the *maximum number of electrons in the valency orbit can not exceed eight.* In order to explain the formation of compounds like phosphorus pentachloride and sulphur hexafluoride, Sugden assumed that some or all the halogen atoms are joined to the central atom *by linkages involving the sharing of a single electrons. This type of union between two atoms by means of single electron is given the name singlet linkage.*

A singlet linkage is formed between atoms one of which has complete octet of electrons and is capable of donating electrons to other atoms each of which is short of one electron. Thus.

"A singlet linkage is like a co-ordinate linkage with the difference that it is formed by the donation of one electron only and not two as in co-ordinate linkage".

In phosphorus pentachloride, three of the chlorine atoms are supposed to be attached to the phosphorus atom by ordinary covalent linkages whereas the other two chlorine atom are joined to it by singlet linkages, thus making a total of eight electron around the nucleus. In order to support his singlet linkages Sugden said that PCl_5 on heating dissociates to give PCl_3 and Cl_2 which indicates that two chlorine atoms in Cl_5 are bonded to phosphorus atom differently. In SF_6, three will be two covalent and four singlet linkages while in OsF_8 there will be eight singlet linkages.

POLARISATION OF IONS OR POLARISING EFFECT

Introduction

When two oppositely charged ions approach each other closely, the cation tends to attract the outer electrons of the anion and at the same time it tends to repel the positively charged nucleus of the anion. This results in the distortion of the anion or in other words the anion is polarised. This is known as polarising effect. The cation is said to cause polarisation and the anion is said to be polarised. The anion also causes a similar effect on the cation.

As the size of the cation is generally smaller than the anion, the polarisation of a cation by the anion occurs to a lesser extent and may be considered to be almost negligible in most case. It means that a cation possesses a strong polarisation power and the anion is more susceptible to polarization, *i.e.*, the anion has more polarisability.

Factors Affecting Polarisation of Ions

The main factors are :

(i) **Charge on cation or anion.** Generally speaking, the polarising power of a cation increases with its charge, the higher the charge on a cation, the greater the polarising power. The reason for this is that a cation having higher charge can attract electrons more effectively. Thus, Ca2+ ion has greater polarising power than Na+ ion.

Similarly, the polarisability of anion increases with its charge, the higher the charge on anion, the greater is its polarisability. The reason is that with an increase in negative charge, the anion will be able to repel its outer electrons more effectively.

(ii) **Size of cation of anion :** The polarising power of a cation increases if its size is smaller while the polarisability of an anion increases if the size is larger.

Importance of Polarisation of Ions

This concept is quite successful in deciding about the nature of bond between two bonded atoms. For instance.

(i) If the polarisation is quite small, the bonding is ionic.

(ii) If the polarisation is very high, the bond is a pure covalent bond..

(iii) If the polarisation is in between the two values expressed in (i) and (ii) the bond is partial covalent bond.

FAJAN'S RULES

Fajan gave a number of empirical rules known as Fajan's rules to explain the polarisation of ions. According to him, the degree of polarisation or distortion of anion by a cation is determined by the following factors :

(a) **Charge on the cation and anion :** As the charge on cation is increased, its tendency to polarise the anion becomes greater and greater, *i.e.*, larger the charge on the cation, greater is the force of attraction between this cation and anion, hence more polarisation. As in the case of cation, charge on anion also affects the polarisation. As the charge on anion increases, it will be more attracted towards cation.

(b) **Size of the ions :** Smaller the cation, greater is the force with which it can attract the outermost electrons of the anions, and larger the anion, lesser is the force with which its nucleus can hold the electrons in its outer shell. Hence, smaller the cation and larger the anion greater is the tendency for the covalency.

(c) **Nature of the solvent :** High dielectric constant of a solvent decreases the force of attraction between the ions, and the ions move away from each other. As a result the tendency to form a covalent bond decreases.

(d) **Electronic configuration of cation :** A cation with 18 electrons in its outermost shell can polarise an anion more effectively than a cation with 8 electrons in its outermost shell. It is because the *d* electrons of the 18 electrons shell shield the nuclear charge less effectively than the s and *p* electrons of the 8 electron shell. *Hence, a cation with 18 electrons shell has larger effective nuclear charge than a cation with 8 electrons in outermost shell, even though the two may have same size and same charge.*

For example. AgCl is more covalent than NaCl, because in the former Ag^+ has 18 electrons in its outermost shell while in the latter Na^+ has 8 electrons in its outermost shell. Size of Ag^+ and Na^+ cations is almost the same.

Fajan originally expressed the rules in terms of the concept of atomic volume, *i.e.*, the atomic weight of an element divided by its specific gravity, this gives an approximate measure of the size of an atom of an element.

It follows from what has been said that an ion will be formed most easily if the element from which the ion is formed has a low atomic volume for an anion and a large atomic volume for a cation. If the conditions prevailing in any particular case do not favour the formation of ions, then a covalent bond will probably be formed rather than an ionic bond.

Applications of Fajan's Rules : Main applications are :

(a) **Ease of formation of ions :** For ions which have an inert gas structure the predominant factors in their formation are the ionic size and the ionic charge. The effect of these two factors on the formation of common cations and anions can be summarised in

Table 3

← Decrease in ionic charge | Decrease in ionic charge →

Cations (↓ Increase in ionic size; ↓ Ions form more easily)	Anions (↑ Ions form more easily; ↑ Increase in ionic size)
Li^{+} Be^{++}	N^{---} O^{--} F^{-}
Na^{+} Mg^{++} Al^{+++}	P^{---} S^{--} Cl^{-}
K^{+} Ca^{++} Sc^{+++}	Br^{-}
Br^{+} Sr^{++} Y^{+++} Zr^{++++}	I^{-}
Cs^{+} Ba^{++} La^{+++} Ce^{++++}	

(b) **Physical properties of chlorides :** The physical properties of compounds like chlorides are affected by the size of cation. As for example the melting point, boiling point, and electrical conductivities in the fused state decrease within a period as shown below in Table 4.

Table 4

	LiCl	*$BeCl_2$*	*BCl_3*	*CCl_4*	*NCl_5*	*OCl_6*
M.P.	606°	404°	– 107°	– 23°	Not	Not
B.P.	1381°	488°	126°	76°	formed	formed
Elec.cond		166	0.06	0	0	

(c) **Diagonal relationships :** The so called diagonal relationships in the periodic table have been shown below.

These are due to the comparative ease of ionisation of elements joined by arrows. Any increase in case of ionisation in passing down a group is counteracted by a decrease in the case of ionisation passing from left to right. Whatever the similarities are between, say lithium and magnesium, they are due to the that they both form ions with the same ease.

(e) **Strengths of hydracides :** Due to the increasing size of anions in sixth and seventh groups, the acidic character increases as we go down the group.

For example, due to the larger size of S^{2-} H_2S is stronger acid than H_2O. Similarly, the acidic character of halogen acids increases with the increase in the size of halide ion *i.e.*, HI is stronger acid than HCl.

LIGATION OR COMPLEXATION ENERGY (E_{tig})

It is the energy released when a gaseous cation combines with a group of gaseous ligands (anions or neutral molecules) to form a gaseous complex as shown by the general equation.

$$M^{n+} (g) + xL^{n-} (g) \rightarrow [MLx]^{x-n} (g).$$

The direct measurement of ligation energy is not possible. However, one can determine it by using a cycle due to Blake and Cotton. This have been illustrated in Fig. 1 for Cs_2CoCl_4, a substance consisting of the crystalline state of Cs^+ ions and $[CoCl_4]^{2-}$ ions.

All of the enthalpies of formation are known, and the lattice energy, U_0 can be computed. Hence the ligation energy (E_{lig}) which is $-\Delta H_{lig}$, can be calculated. The value of ΔH_{lig} is about 2100 kJ, although, values in the range of 2016 to 2310 kJ/mole are possible.

$$2Cs^+(g) + Co^{2+}(g) + 4Cl^- \xrightarrow{-\Delta H_{lig}\ (=E_{lig})} 2Cs^+(g) + [CoCl]^{2-} (g)$$

ΔH_f of Co^{2+} (g) $\qquad -U_0$

$2\Delta H_f$ of Cs^+ (g)

$4\ \Delta H_f$ of Cl^- (g)

$$2Cs\ (S) + 2Cl_2 + Co\ (S) \xrightarrow{\Delta Hf} Cs_2CoCl_4\ (S)$$

A thermochemical cycle which includes the energy of formation of gaseous $[CoCl_4]^{2-}$ from gaseous Co^{2+} and Cl^- ions.

ODD ELECTRON BONDS

Introduction

In 1916. *W.F. Luder* postulated that there are a number of stable molecule in which the bonds are formed by sharing of an odd number of electrons (*i.e.*, one, three, five, etc) between the two bonded atoms. *The bonds of this type are called odd electrons bonds.*

Types : Odd electrons bonds are of following types.

(a) **One electron bond :** If a bond is formed by sharing of one electron between the two bonded atoms, it is known as one electron bond.

(b) **Three electron bond :** If a bond is formed by sharing of three electrons between the two bonded atoms, it is termed as a three electron bond.

(c) **Five electron bond :** If five electrons are shared between two bonded atoms in a molecule, it is called a five electron bond.

Similarly, one can define seven, nine, etc. electron bonds. However, we will restrict our discussion to one electron bond and three electron bond only.

Example of One Electron Bond

An outstanding example of one electron bond is found in the hydrogen molecule ion, H_2^+. This ion has been detected spectroscopically. This ion is characterised by a bond energy of 256 kJ per mole and on the other hand hydrogen molecule, H_2, has a bond energy equal to 457.8 kJ per mole. The hydrogen molecule ion may be considered to be a combination of the two resonance forms. Both the forms are of equal stability (*i.e.* equal energy).

$$^{+}H\ .\ H \leftrightarrow H\ .\ H^{+}.$$

Internuclear distance between the hydrogen atoms in H^+ ion is 1.06Å and in H_2 it is 1.06Å

One electron bond in H_2^+ ion is weaker than the normal covalent bond in H_2 molecule due to the 1 following reasons :

(i) In hydrogen molecule two electrons are of opposite spin and thus they have less repulsive nature than the two electrons of the same spin and hence causing the H_2 molecule to be stable. In H^+ ion there is only one electron which is approximately half as effective in reducing the mutual repulsion of positively charged nuclei. This causes the less stability in H^+ ion.

(ii) As the internuclear distance in $H|^+$ ion (1.06Å) is longer than two electron bond in H_2 (0.74Å), it means than one electron bond is weaker than the normal covalent bond.

Linnett utilised the above typical relationship of bond length to bond strength to postulate the existence of one electron bond in nitrosyl halide. As the N–Cl bond length in nitrosyl chloride is greater than in $NHCl_2$ molecule (N–Cl bond length in NOCl is 1.06 Å and N–Cl bond length in $NHCl_2$ is 1.76Å). Linnett proposed that nitrogen-chlorine bond in NOCl is a one electron bond and that in $NHCl_2$ is a normal covalent bond.

The excited states of alkali metal molecules involve odd electron bonds. Thus, L_2^+, K_2^+, N_2^+ are all bound by one electron bond. Bond energies indicate that they are 43% relatively weaker in bond strength to their neutral molecular species.

Examples of Three Electron Bond

If three bonding electrons are available, the following are the two structures involving two atoms A and B without the violation of Pauli's principle :

A :.B A .: B
(a) (b)

Under normal conditions both structures (a) and (b) are not feasible. However, the two structures would be energetically feasible if A and B are sufficiently alike; the two structures might then possess essentially the same energies and resonance between them might stabilise a bond which involves three electrons, *i.e.*, A . . . B. Some examples of three electron bond are discussed below :

(i) **He_2^+ ion :** This ion may be represented as a resonance hybrid of (a) and (b).

$$\underset{(a)}{He \overset{x}{\underset{x}{}} He^+} \rightleftarrows {}^+He \quad \underset{(b)}{{}^+He \overset{x}{\underset{x}{}} He}$$

Due to the resonance between the two forms (a) and (b), there results in the formation of a three electron bond between He and He^+, Thus, He_2^+ ion is a He ....He. The binding energy in He_2^+ ion is 243 kJ per mole. The three electrons involved in the three electron bond in He_2^+ ion come from similar atomic orbitals in the isolated atoms.

(ii) O_2 molecule : The normal valence bond treatment gave the following structure of oxygen molecule

$$:\underset{..}{O}:\overset{x}{\underset{x}{}}\underset{xx}{O}\overset{x}{\underset{x}{}} \quad \text{or} \quad :\underset{..}{O}=\underset{xx}{O}\overset{x}{\underset{x}{}}$$

The above electronic structure with no unpaired electrons fails to account for the paramagnetic behaviour of oxygen although it accounts for high dissociation energy of oxygen molecule. Another structure involving three electron bonds has been suggested by Pauling which accounts for two unpaired electrons and also for the high energy of dissociation.

$$:O \overset{\cdots}{\underset{\cdots}{-}} O:$$

Evidently, the oxygen atoms are joined by one electron pair bond (*i.e.*, one covalent bond) and two three electron bonds. Thus, the three electron bonds have one unpaired electron each. Such a molecule is known as a *" double odd molecule".*

(iii) Nitric oxide molecule : This molecule has two normal covalent bonds and one three electron bond. Thus, it may be represented as.

$$:N \overset{.xx}{\underset{t}{=\!=}} O\overset{x}{\underset{x}{}}$$

There are four valence orbitals of each atom, *i.e.*, 2s, $2p_x$, $2p_y$, and $2p_z$. Out of these, one is used for the lone pair of electrons, two for the double bond electrons and the fourth is used for the three electron bond This molecule does not have any tendency to dimerise.

(iv) Nitrogen dioxide : This molecule has a strong tendency to dimerise. The two resonating extreme structures for this molecule are.

(v) OF molecules : It possesses the following structure.

$$:\ddot{O} \overset{\cdots}{-} \ddot{F}:$$

(vi) Chlorine dioxide molecule : The two resonating extreme structures for this molecules are.

The bond lengths are appreciably shorter than for single Cl–O bond. The observed Cl–O bond length in ClO_2 molecule has been found to be 1.49Å and the bond angle $O\hat{Cl}O$ is 116.5°.

(vii) **Superoxide ion (O_2)** : It contains an odd-electron bond and its structure may be written as follows.

$$\left(:\underset{\cdot\cdot}{O}\overset{\cdots}{-}\underset{\cdot\cdot}{O}:\right)^{-}$$

The above structure of superoxide ion has been verified by the measurement of the magnetic susceptibility of potassium superoxide (KO_2). The interatomic oxygen-oxygen distance has been found to be 1.28Å which has been found to be in agreement with that expected for single O–O bond plus a three-electron bond.

Properties of Odd Electron Bonds

(i) The odd electron bonds are generally established either between two like atoms or between different atoms which should not differ by more then 0.5 in electronegativity values because only then the resonance structures of almost equal stabilities would be realised.

Formation of one electron bond is generally occurring between like atoms whereas its formation between unlike atoms is very rate.

(ii) Odd electron bonds are approximately half as strong as a normal covalent bond. For example, the three electron bond is about half as strong as a covalent bond.

(iii) Molecules containing odd electrons are extremely reactive and have the tendency to dimerise. For example, the energy of dimer of molecule containing three electron bonds will be close to that of two monomeric molecules.

$$A\overset{\cdots}{-}B + A\overset{\cdots}{-}B \xrightarrow{\text{Dimerisation}} :A-B-B-A:$$

(iv) Bond length of one electron bond is greater than that of a normal covalent bond whereas the bond length of a three electron bond is intermediate between those of a double and a triple bond. For example, the bond length of a three electron bond in nitric oxide (1.15Å) is intermediate between the bond length for a double bond 1.20Å) and for a triple bond (1.06Å).

(v) One electron bond between two atoms A and B is generally considered to be a resonance hybrid between the following two structures :

$$A.\ B \longleftrightarrow A\ .B$$

Similarly, a three electron bond may be considered to be resonance hybrid of the following two structures.

$$A \bullet \text{I}\ B \longleftrightarrow A : \bullet B$$

Molecular Orbital Theory (MOT)—Treatment of Odd Electron Bonds : When the **MOT** is applied to a one electron bond, it is observed that one of the bonding molecular orbital is half-filled. Thus molecular orbital configuration of H_2^+ ion is given as, :

$$(\sigma_{1s})^1$$

When the **MOT** is applied to a three electron bond, it is found that one of the bonding molecular orbital is completely filled and an antibonding molecular orbital is half-filled. For example, the molecular orbital configuration of NO molecule is

$$N\ [1s^2, 2s^2, 2p_x^1, 2p_y^1 . 2p_z^1] + O\ [1s^2, 2s^2, 2p_x^1, 2p_y^1 . 2p_z^1] \rightarrow$$

$$N\ O\ [K\ K\ (\sigma_{2s})^2, (\sigma_{2s}^*)^2\ (\sigma_z p_z)^2\ (\pi_z p_z)^2\ (\pi_x p_x)^2\ (\pi_x^* p_x)^1]$$

In this configuration $(\pi_x p_x)^2\ (\pi_x\ p_x)^1$ represents a three electron bond in No molecule.

RESONANCE

Concept : The term resonance refers to the fact that *many molecules cannot be represented by any single valency bond structure, but must be considered as indeterminate between two or more such structures. Such molecules are called resonance hybrids.* This term does not imply the molecule resonances (in the physical sense) between the structures; the molecule is not well represented by one structure during part of its life time, by another during another part ; rather the structure is unique, but is most conveniently described by the mathematical expedient called resonance.

Thus, benzene, for example, cannot be represented by any one of the structures I to V alone but intermediate between all of them, and many others resembling structures given below : The canonical forms have no physical significance because these are merely a convenient way of picturing a molecule in terms of familiar Lewis structures. Mathematically this is expressed by saying that the wave function describing the molecule, Ψ_G, is a linear combination of the wave functions, Ψ_1, Ψ_2, Ψ_3... describing the various strictures.

$$\Psi_G = C_1 \Psi_1 + C_2 \Psi_2 + ...$$

The classic difficulty over the concept of resonance is the disposition to regard the molecule as oscillating from one to another of the assigned forms. The fault is not with the molecule but rather with the human inability to describe properly this hybrid by the single structural formula that characterises it.

Bond Order for Resonance Systems : It is very easy to calculate the bond order for molecules or ions (*i.e.*, CO_3^{2-}, NO^{3-}, etc.) having symmetrical canonical structures. For example, the C–O bond order in CO_3^{2-} comes out to be $1\frac{1}{3}$ because its various canonical forms (a) to (c) are contributing equally and each oxygen atom is carrying a negative charge of 2/3 units. Similarly, the C–O bond order in a carboxylate ion comes out to be 1½ and each oxygen carrying a negative charge of 1/2 unit.

The C–O bond order in CO_2 has been found to be greater than 2 because of the non-equivalence of its various canonical forms.

$$\underset{(a)}{O=C=O} \longleftrightarrow \underset{(b)}{\overset{-}{O}-C\equiv\overset{+}{O}} \longleftrightarrow \overset{+}{O}\equiv C-\overset{-}{O}$$

Conditions Necessary for Resonance

If Ψ_1. Ψ_2. Ψ_3... are the various wave functions of the possible structures (*i.e.*, the resonance forms) of a molecule contributing to its resonance hybrid (*i.e.*, the actual structure), the important ones will be those for which the C is large in the following expression.

$$\Psi = N\,(C_1 \Psi_1 + C_2 \Psi_2 + C_3 \Psi_3 + ...)$$

where Y represents the wave function of the actual structure of the molecule. It must be emphasized that Ψ represents a wave function which is the sum of the wave functions Ψ_1, Ψ_2, Ψ3... These various wave functions are included in the resultant (Ψ) but they are not individually characteristics.

1. *The location of the various atoms in each of the canonical structures must be same as in the actual molecule (resonance*

hybrid) but the arrangement of electrons may differ. For example, the two structures of N_2O molecule having the skeletons N–N–O and N–O–N cannot be regarded as the canonical (resonating) structures because the relative positions of the atoms N, N and O are not the same in the two skeletons. Similarly, H–O–Cl and H–Cl–O cannot be canonical structures of HClO because the relative positions of HCl and O are not the same in the two structures.

2. *Contributing canonical (resonating) structures must have the same number of unpaired electrons so that a continuous change from one bond type to another may occur.* For example, structures I and II are not the resonating structures of oxygen because the number of unpaired electrons in the two structures is different

$$\underset{\text{I}}{:\ddot{O}:\overset{xx}{\underset{x}{\overset{x}{}}O}\overset{x}{\underset{x}{}}} \quad \text{and} \quad \underset{\text{II}}{:\underset{\cdot}{\ddot{O}}\;{}^{\times}\underset{xx}{\overset{x}{O}}\overset{x}{\underset{x}{}}}$$

3. *Distribution of charges in a contributing structure must be reasonable.* This rule may be considered in two parts.

(a) Like charges should not reside on adjacent atoms in a resonating structure otherwise the structure will appear to be unstable as a result of the electrostatic repulsion. For example, a structure such as $\overset{-}{A} - \overset{+}{B} - \overset{+}{C} - \overset{-}{D}$ is therefore unlike to play a major role in hybrid formation.

(b) Unlike charges should reside on atoms close together in a contributing form. Furthermore, placement of adjacent charges of opposite sign will be more favourable than when these are separated.

The above mentioned two conditions (a) and can be understood by considering the contributing structures (I) and (II).

$$\underset{\text{I}}{H - \ddot{N} = \overset{+}{N} = \overset{-}{N}} \longleftrightarrow \underset{\text{II}}{H - \overset{-}{\underset{\cdot\cdot}{N}} = \overset{+}{N} \equiv N:} \longleftrightarrow \underset{\text{III}}{H - \overset{+}{N} \equiv \overset{+}{N} - \overset{2-}{\underset{\cdot\cdot}{\ddot{N}}}:}$$

The contributing structure III is unimportant because positive charges reside on two adjacent nitrogen atoms and also because of the double negative charge on nitrogen atom.

4. The positive charge should reside as far as possible on the less electronegative element and *vice versa.* For example, in HF, the structure H^+F^- is important while the structure $H^- - F^+$ has high energy and is unimportant.

$$H - F \leftrightarrow H^+F^- \longleftrightarrow F^+$$

5. *The proposed canonical structure should have a maximum number of bonds.* It means that greater the number of bonds, the greater is the importance of the canonical structure. Let us consider BF. Which is a resonance hybrid of the following four contributing structures.

   Out of the four structures, three structures II, III and IV contribute more to the actual structures because each of these contains one additional bond, even though a positive formal charge is acquired by a fluorine atom.

6. *Contributing canonical structures should posses almost identical energy.* When they have equal energy, they contribute equally. For example, the three resonance forms of carbonate ion contribute equally because the energies of the three contributing structure are identical.

In many cases, however, the energies of the structures differ and in these cases we find that the contribution of a canonical structure is inversely proportional to its energy.

It means that the canonical forms having higher energies are unstable. For example, hydrogen molecule is a resonance hybrid of the following four structures.

$$\underbrace{\overset{1}{H_a} - \overset{2}{H_b}}_{\text{I}} \quad \underbrace{\overset{2}{H_a} - \overset{1}{H_b}}_{\text{II}} \quad \underbrace{\overset{1,2}{\bar{H}_a} \; \overset{+}{H_b}}_{\text{III}} \quad \underbrace{\overset{+}{H_a} - \overset{1,2}{\bar{H}_b}}_{\text{IV}}$$

Covalent Forms (I, II) — Ionic Forms (III, IV)

Structures I and II are covalent forms in which two hydrogen atoms are held by covalent bonds. Structures III and IV are ionic forms in which both the electrons are associated with the Ha (III) or Hb (IV) respectively.

Out of these four structures, the ionic structures are relatively unstable because they possess higher energy and hence, they make a small contribution (about 2%) to the actual structure of H_2 molecule.

A less obvious result is that when contributing structures differ in bond angle, resonance will be reduced. Consider, for example, the following resonance structures for nitrous oxide.

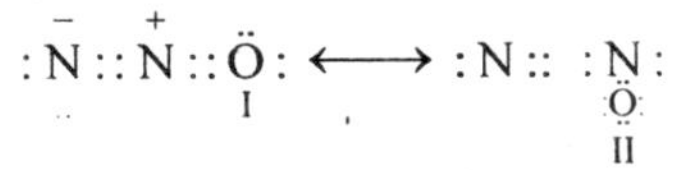

$$\underset{\cdot\cdot}{\overset{-}{:N}}::\overset{+}{N}::\ddot{O}: \longleftrightarrow :N::\ :N:$$

I, II (structure II: N::N with O bonded below the second N, forming a ring)

Aside from the fact that structure II is a strained structure and therefore less stable than I, it will not contribute to the resonance of N_2O since the bond angle is 183° in I and 0° in II. For any intermediate hybrid, the contribution of either I or II could be unfavourable because of the high energy loss when I is bent or when II is opened up.

Properties of Resonance Hybrid

(i) A resonance hybrid cannot be expressed by any single structure.

(ii) The stability of resonance hybrid is greater than that of the contributing structures.

The observed heat of combustion of benzene is 150.6 kJ mol^{-1} which is lower than the value calculated for any of the resonating (canonical) forms. This difference in energy is known as *resonance energy*. The greater the resonance energy, the greater is the stabilization due to resonance.

This is also explained by the heat of formation of the compound. The observed heat of formation of carbon dioxide is 1590 kJ mol^{-1} which is 125.5 kJ more than the calculated value. In other words carbon dioxide requires 125.25 kJ more energy than expected to break it up into elements *i.e.* O = C = O. The difference in heats of formation, observed and calculated is also known as *resonance energy*.

Resonance Energy

One of the most important points connected with resonance is that the resonance hybrid is a more stable structure than any of the structures contributing to it.

The increased stability of a resonance hybrid than that of any individual canonical structure is accounted in the following way : If it is considered that a resonance hybrid has two resonating structures, I and II, they can each be represented by wave functions, in simple cases, which give the energies of the structures they represent. By combining the two wave functions it is found that a third wave function will also represent the system and that this function corresponds to a lower energy value, *i.e.* a higher stability.

The difference between the energy content of the resonance hybrid and other canonical structure having least energy is called resonance energy or delocalization energy. Resonance energy can be calculated from heat of hydrogenation and heat of combustion values.

When we say that the resonance energy of carbon dioxide is 159.5 kJ/mole, it means that the actual carbon dioxide molecule is more stable than the theoretically calculated stability of the carbon dioxide with the conventional Lewis formulation by 159.5 kJ/mole.

The source of resonance energy is the same as that of any other type of bonding energy. As the energy of a single or a double bond arises from the ability of an electron to encompass two nuclei in a molecule instead of one in the case of an atom.

Similarly resonance energy is merely an additional stabilisation resulting from the ability of an electron (π – or unshared) to encompass three or more nuclei instead of two. These electrons, encompassing more than two nuclei, are frequency called mobile electrons and are, therefore, said to be delocalized electrons in the σ-bonds. Therefore, resonance energy is also called *delocalization energy.*

In case of many organic molecules resonance energy can be calculated from heats of formation and/or heats of combustion. When the heat of formation from the elements of an organic compound is greater than calculated, or the heat of combustion is lower than calculated, the molecule is stabilized by the resonance.

The difference between observed and calculated values is the resonance energy. The greater the resonance energy, the greater is the stabilization due to resonance. The resonance energy is maximum when the canonical structures have equal the contributing structures are many.

Examples of Molecules Showing Resonance

1. **Carbon monoxide molecule, CO :** Both carbon as well as oxygen atoms are having two unpaired electrons. Thus carbon monoxide must be having C = O structures. But following experimental results do not favour this structure.

(a) **Bond length :** The actual carbon-oxygen bond length (1.13 Å) in carbon monoxide has been shorter than double bond (1.22 Å) but longer than triple bond (1.10Å).

(b) **Bond energy :** The actual bond energy (1057 kJ/mole) has been found to be greater than calculated bond energy (726 kJ/mole).

(c) **Dipole moment :** The experimental dipole moment of C = O has been negligibly small. However, on the basis of higher electronegativity (3.5) of oxygen than carbon (2.5), the calculated dipole moment would be appreciable.

Keeping in view the above discrepancies, the following canonical forms of carbon monoxide are postulated.

$$\underset{\text{I}}{:C::\ddot{O}:} \longleftrightarrow \underset{\text{II}}{:\overset{+}{C}:\overset{-}{\underset{..}{\ddot{O}}}:} \longleftrightarrow \underset{\text{III}}{:\overset{-}{C}:::\overset{+}{O}:}$$

Structure I is covalent whereas structures II and III are ionic. When the C–O bond length is determined experimentally, it is found to be 1.13Å. Thus shown that the bond between carbon and oxygen is triple bond. Hence the structures I and II make relatively smaller contributions.

2. **Carbon dioxide molecule CO_2 :** Carbon atom is having four unpaired electrons in its excited state ($1s^2\ 2s^1\ 2p_x^{\ 1}\ 2p_y^{\ 1}\ 2p_z^{\ 1}$). Oxygen atom is having two unpaired electrons ($1s^2\ 2s^2\ 2p_x^{\ 2}\ 2p_z^{\ 1}$). Thus carbon dioxide might be O = C = O structure. But following experimental results do not support this structure.

(a) **Bond length :** The actual carbon-oxygen bond length (1.15 Å) in CO_2 has been found to be shorter than double bond (1.22 Å) but longer than triple bond (1.10 Å).

(b) **Bond energy :** The actual bond energy has been 150 kJ per mole. This is more than calculated bond energy.

Keeping in view the above discrepancies, the following three canonical forms of CO_2 are suggested

$$\underset{\text{(I)}}{O = C = O} \leftrightarrow \underset{\text{(III)}}{\overset{-}{O} - C \equiv \overset{+}{O}} \leftrightarrow \underset{\text{(III)}}{\overset{+}{O} - C - \overset{-}{O}}$$

Structure (I) is covalent whereas the remaining two structures (II) and (II) are ionic. All the three structures make nearly the same contribution. It is more stable than any of the three reasonating structures.

3. Oxygen molecule : According to valence bond theory, oxygen molecule is having the structure, FIGURE According to this structure, there exists a double bond between two oxygen atoms. Oxygen will have the following resonating structures.

$$:\ddot{O}=\ddot{O}: \longleftrightarrow :\overset{+}{\underset{..}{O}}-\overset{-}{\underset{..}{\ddot{O}}}: \longrightarrow :\overset{-}{\underset{..}{\ddot{O}}}-\overset{+}{\underset{..}{O}}:$$

Oxygen molecule is paramagnetic (attracted by magnetic field) and paramagnetism has been ascribed to one or more unpaired electrons. All the above structures are not having any unpaired electrons and hence are unable to explain paramagnetic character of oxygen. In order to account for the paramagnetic behaviour of oxygen. Pauling postulated the following structure for oxygen molecule.

$$:O\underset{X..}{\overset{X...}{\text{——}}}O:$$

This structure possesses one covalent bond and two 3-electron bonds. Hence oxygen molecule is having two unpaired electrons (×, ×) which account for its paramagnetic nature. However, this structure fails to obey the octet rule. The paramagnetic behaviour has been successfully explained by molecular orbitals theory.

4. Carbonate ion (CO_3^{2}) : According to valence bond theory, the carbonate ion is having the following structure.

$$\bar{O}-\overset{\overset{O}{\|}}{C}-\bar{O}$$

In this planar structure, carbon atom gets bonded to

(a) Two oxygen atoms by single covalent bond and,

(b) One oxygen atom by double covalent bond.

It means that all the carbon oxygen bonds have been not of equal length. But X-rays studies reveal that all the carbon oxygen bonds have been of equal strength. Pauling explained the equivalence of all C–O bonds by postulating the following resonating structures.

5. Benzene (C_6H_6) : The various structures of benzene are depicted.

Structures I and II have been due to Kekule. Structures III, IV and V have been due to Dewar and structure VI has been such that two carbon atoms have only seven electrons (*i.e.*, one electron less than eight electrons required by octet rule). Structure VI has been thus

$$:N\underset{t}{\overset{.xx}{=\!=}}O\overset{x}{\underset{x}{}}$$

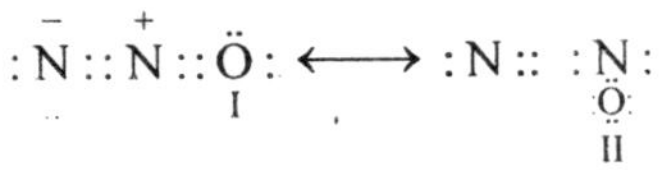

$$\underset{\text{I}}{:\overset{-}{\underset{..}{N}}::\overset{+}{N}::\ddot{O}:} \longleftrightarrow \underset{\text{II}}{:N::\ :\underset{\substack{\cdot\cdot\\ \ddot{O}:}}{N}:}$$

$$:\ddot{O}=\ddot{O}: \longleftrightarrow :\overset{+}{\underset{\cdot\cdot}{O}}-\overset{-}{\underset{\cdot\cdot}{\ddot{O}}}: \longrightarrow :\overset{-}{\underset{\cdot\cdot}{\ddot{O}}}-\overset{+}{\underset{\cdot\cdot}{O}}:$$

$$:O\overset{X\ldots}{\underset{X..}{\text{---}}}O:$$

unimportant and hence is making least contribution. Structures III to V are also less important due to the presence of long C–C bonds as compared to other C–C bonds. The structures I and II are real resonating structures because these are only able to explain the equivalence of all C–C bond length has been in between C–C and C = C bond.

6. **SO_3** : It affords another prominent example of a molecule having resonance hybrid structures. Its various contributing structures are as follows.

7. **N_2** : As nitrogen molecule has the same number of electrons, *i.e.*, 10 as in carbon monoxide, it will have the resonating structures as that of carbon monoxide.

$$\underset{\text{I}}{:N:::N:} \longleftrightarrow \underset{\text{II}}{:\overset{-}{\ddot{N}}::\overset{+}{N}:} \longleftrightarrow \underset{\text{III}}{:\overset{+}{N}::\overset{-}{\ddot{N}}:}$$

8. **H_2** : It is a resonance hybrid of the following structures.

$$\underbrace{\overset{1}{H_a} - \overset{2}{H_b} \quad \overset{2}{H_a} - \overset{1}{H_b}}_{\substack{\text{I} \\ \text{Covalent Forms}}} \qquad \underbrace{\underset{\text{III}}{\overset{-}{Ha}\;\overset{+}{\overset{1,2}{Hb}}} \quad \underset{\text{IV}}{\overset{+}{Ha}\;\overset{-}{\overset{1,2}{Hb}}}}_{\text{Ionic Forms}}$$

Structures I and II are covalent forms in which two hydrogen atoms are held by covalent bonds. Structures III and IV are ionic forms in which both the electrons are associated with the Ha (III) or Hb (IV) respectively. Out of these four structures, the ionic structures are relatively unstable because they possess higher energy and hence, they make a small contribution (about 2%) to the actual structure of H_2 molecule.

9. **Nitric oxide molecule, NO** : It is considered to be resonance hybrid of the structures given below.

$$\overset{\times}{\underset{\times}{\times}}N\overset{\times}{\underset{\times}{}} \; :\ddot{O}: \longleftrightarrow \overset{+}{\overset{.\times}{\underset{\times}{\times}N\overset{\times}{\underset{\times}{}}}} \; :\overset{+}{\dot{O}}: \\ \quad - \quad -$$

or

$$\underset{\text{I}}{:\dot{N} = \ddot{O}:} \longleftrightarrow \underset{\text{II}}{\overset{\times .}{N} =: \dot{O}:}$$

From structures I and II, it follows that nitric oxide molecules is an odd–electron molecule.

$$\underset{\text{(III)}}{\overset{\times}{\underset{\times}{}}\dot{N} = \ddot{O}:}$$

The above structure has one three-electron and two electron (*i.e.*, one double bond) bonds between N and O. The structure III suggests a small dipole moment. Also the experimental value of dipole moment, is 0.16 D which is a small value. The calculated values of bond length for N = O and N ° O are 1.15Å and 1.0Å respectively, whereas the observed value of bond length in nitric oxide is 1.04Å.

10. Dinitrogen oxide (Nitrous oxide) N_2O : It is supposed to be a resonance hybrid of the structures shown below.

$$\underset{(I)}{N=N=O} \longleftrightarrow \underset{(II)}{N\equiv N} \longrightarrow O$$

The calculated bond lengths for the various bonds involved in these structures are N = N, 1.2Å N ° N, 1.1Å. N = O, 1.15Å, and N–O, 1.36Å. The length of the molecule, which is linear, is 2.31Å. The probable bond lengths being N–N, 1.12Å and N–O, 1.19Å. The dipole moment of nitrous oxide is very small so that the two structures must contribute almost equally.

11. Sulphate ion, SO_4^{2-} : In the sulphate ion, the four oxygen atoms are arranged around the central sulphur atom almost tetrahedrally. The measured bond distances are all equal to 1.51 Å. The calculated bond distance for single bond between sulphur and oxygen atoms is 1.7Å, and for double bonds, 1.49Å.

It is clear that there must be some double bond character about the bonds in the ion, and is best represented as a resonance hybrid between many structures of which some typical ones are shown as follows.

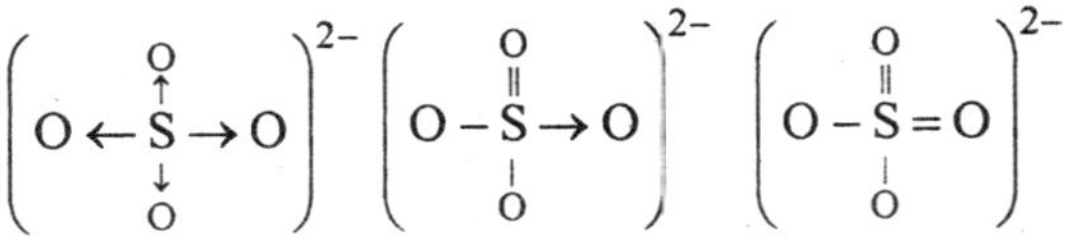

$$\left(\begin{matrix} & O & \\ & \uparrow & \\ O\leftarrow & S & \rightarrow O \\ & \downarrow & \\ & O & \end{matrix}\right)^{2-} \left(\begin{matrix} & O & \\ & \| & \\ O- & S & \rightarrow O \\ & | & \\ & O & \end{matrix}\right)^{2-} \left(\begin{matrix} & O & \\ & \| & \\ O- & S & =O \\ & | & \\ & O & \end{matrix}\right)^{2-}$$

The last of these structures is probably the most important but there are many similar ones because the double bond may have alternating positions.

12. Hydrazoic acid, HN_3 : The azide ion has a linear symmetrical structure. In hydrazoic acid, NH_3, however, the symmetry is lost and the molecule has a bent structure and is a resonance hybrid as shown b below.

$$H:\ddot{N}::\overset{+}{N}::\overset{-}{\ddot{N}}: \longleftrightarrow H:\underset{\cdot\cdot}{\ddot{N}}:\overset{+}{N}:::N:$$

13. **Sulphur dioxide, SO_2 :** It is a resonance hybrid of the following structures.

The experimental values for two S–O bonds are of equal length 1.43 Å and have some double bond character.

14. **Nitrite ion, NO_2 :** It is isoelectronic with SO_2 and is believed to exist in resonance forms similar to sulphur dioxide.
15. **Carbonate ion, CO_3^{2} :** It is a resonance hybrid among the following structures.

The calculate bond length for the C = O bond is 1.21Å and for C = O bond. 1.43°. The observed bond length in the carbonate ion is 1.31Å. The resonance energy is 176 kJ mole^{-1}.

16. **Nitrate (V) ion, NO_3 :** It is a resonance hybrid among the following structures.

The observed bond length is 1.21Å. The calculated bond length for N – O, is 1.36Å and for N = O. 1.15Å. The resonance energy is 188.3 kJ mole^{-1}.

17. **Boron trifluoride, BF_3 :** All the B–F bonds in BF_3 are of equal length and possess some double bond character. This observation can be explained by assuming the molecule to be a resonance hybrid of the following three structures.

Each B–F bond possesses some double bond character due to the presence of pπ–pπ bonding,

The effect of resonance on heat of formation and bond lengths. Resonance in a molecule produces the following two results.

(i) The stability of the resonance hybrid is greater and its heat of formation is therefore lower than those of any of the resonance structures. Hence, the observed heat of formation of the molecule is lower than those calculated for the resonance structures.

(ii) The lower the heat of formation of a molecule, the greater is the energy which has to be supplied to disrupt its bonds, *i.e.*, the greater the strength of the bonds. The increased strength of the bonds in resonance hybrid results in a shortening of the average bond lengths. Hence the observed bond lengths for those bonds where the electrons are delocalized are shorter than the average bond lengths of the corresponding bonds in the resonance structures.

For example, the normal bond length of carbon-oxygen double bond is 1.22Å while in carbon dioxide it is actually 1.15Å. Also in Kekule structure for benzene there are three carbon-carbon double bonds (1.34 Å) and three carbon-carbon single bonds (1.54Å) and all the six carbon-carbon bonds in benzene are actually equal and the bond length in each case is 1.39Å– intermediate between carbon–carbon single bond and carbon-carbon double bond.

HYDROGEN BONDING

Many compounds are known in which the hydrogen atom exists simultaneously between two strongly electronegative atoms and thus acts as a bridge between them. In this situation, the hydrogen atom is involved in two bonds : one is *natural covalent bond* and the' second is known as *hydrogen bond.* The hydrogen bonding is illustrated below for the association of several molecules of hydrogen fluoride.

$$\overset{+\delta}{H} - \overset{-\delta}{F} \ldots\ldots\ldots \overset{+\delta}{H} - \overset{-\delta}{F} \ldots\ldots \overset{+\delta}{H} - \overset{-\delta}{F}$$

The hydrogen bond is generally represented by a dotted line as shown above. The solid line represents the original covalent bond.

Cause of Formation of Hydrogen Bond

A hydrogen atom contains one proton and one electron. When the hydrogen atom is attached to a strongly electronegative atom (say, F), the electrons of the covalent bond are displaced towards the electronegative atom F. This consequently results in the development of a fractional positive charge on the hydrogen atom, and the fractional negative charge on the electronegative atom, F.

$$\overset{+\delta}{H} - \overset{-\delta}{F}$$

When two H–F molecules approach each other the H will exert a strong electrostatic force and so can attract F of other molecule. The attractive force between hydrogen and F leads to the formation of a weak bond called hydrogen bridge or hydrogen bond.

$$\overset{+\delta}{H} - \overset{-\delta}{F} \ldots\ldots \overset{+\delta}{H} - \overset{-\delta}{F} \ldots\ldots \overset{+\delta}{H} - \overset{-\delta}{F} \ldots\ldots \overset{+\delta}{H} - \overset{-\delta}{F} \ldots\ldots$$

Hydrogen bond is formed with such electronegative atoms which are having small cationic radii. Fluorine, oxygen, and nitrogen are of this character.

Atom bigger in size than, N, O or F are not able to form H-bond. *The reason for this may be put as follows.*

Atoms bigger in size than N, O or F are not able to attract covalent electrons effectively. It occurs due to the screening effect of electrons between valence electrons and nucleus. Thus, the covalent bond between hydrogen and other atom would be less polar. The electrostatic force of attraction becomes very weak in them. Therefore, H–bond does not get formed, *e.g.*, H–bonding in HCl or HBr would not be due to big size of Cl and Br–atoms. Hence, the factors favouring the H–bond include :

(i) High electronegativity, and

(ii) Small size of atom bonded to H–atom.

Types of Hydrogen Bonds

There are two types of hydrogen bonds :

(a) Intra-molecular hydrogen bonding : *This type of bond arises due to the electrostatic forces of attraction between hydrogen and an electronegative element present in the same molecule. For example.*

(i) The compound salicylaldehyde contains a hydroxyl and aldehyde group adjacent to each other thereby permitting the formation of a hydrogen bond between two groups.

(ii) Other compounds in which such intramolecular hydrogen bonding takes place are o-chlorophenol, o-hydroxy benzoic acid, enolic form of acetoacetic ester. The enolic-form of acetoacetic ester has lower boiling point than the keto form. This is probably due to intra-molar hydrogen bonding in the enol form which prevents intermolecular bonding and thus prevents association which raises the boiling point.

Conditions for intra-molecular hydrogen bonding : In case of intermolecular case hydrogen bonding, the most important condition is that the molecules must contain one hydrogen atom linked to one highly electronegative atom.

In case of intra– molecular hydrogen bonding the following conditions are favourable for hydrogen bonding.

(i) The molecule should contain two groups so that one group contains H–atom linked to highly electronegative atom and the other group should also contain a highly electronegative atom linked to a lesser electronegative atom.

(ii) The molecule should be planar.

(iii) The hydrogen bonding should lead to the formation of at least a six–membered ring including the H–atom.

(b) Inter molecular hydrogen bonding : *When the hydrogen bond is formed between the different molecules, the bonding is known as inter-molecular hydrogen bonding. Some examples are.*

(i) Crystalline hydrogen fluoride consists of infinite chains of the type as shown with an H–F–H angle of 120°. These chains persist to a large extent in the liquid and gaseous phases.

As might be expected each hydrogen atom lies closer to the more firmly attached, covalently-bonded fluorine atom than to the hydrogen-bonded fluorine atom. Heating progressively shortens the length of the chains until above the boiling point, the associated units, $(HF)_x$ become quiet, small, x can be any integer upto 6. In aqueous solution hydrogen fluoride ionises to yield not the simple F– ion, but the bifluoride ion, HF_2^- (giving salts of the type KHF_2). The HF^-_2 ion is a resonance hybrid for which the following resonance structures can be written as follows.

$$^-F - H - F \leftrightarrow F - H - F^-$$

A consequence of this resonance is that the hydrogen atom is equidistant from the two fluorine nuclei. This is exceptional for hydrogen bonding.

(ii) Alcohols form associated molecules like water molecules.

```
 R      R      R                 H      H      H
 |      |      |                 |      |      |
H–O....H–O....H–O....          H–O–H....OH....– O....
```

The hydrogen bonding or the association of molecules may extend to several molecules.

(iii) The formation of hydrogen bonds between molecules of HCN is made possible by the triple bond which enhances the electronegativity of the nitrogen atom and of the carbon, just as the triple bond in acetylene enhances the electronegativity of the two carbon atoms. Hence, the polymerisation occurs.

$$H - C ^\circ N \ldots H - C \equiv N \ldots H - C \equiv N \ldots$$

(iv) Gentle heating of $CuSO_45H_2O$ gives $CuSO_4.H_2O$ and at much higher temperature the last molecule of H_2O is removed. This suggests that one of the molecules of water of crystallisation is different from other four, and leads to the formula. $Cu(H_2O_4)$ SO_4. H_2O.

In its crystal structure, each Cu^{2+} ion is surrounded octahedrally by six O-atoms, four of which are of four water molecules and the two are of SO_4^{2-} ion.

Thus, four water molecules are associated with Cu^{2+} ion as coordinated water and the remaining fifth water molecule is held to SO_4^{2-} ion and to two of the four coordinated water molecules by hydrogen bonds to give an overall structure to $Cu(H_2O)_4$ SO_4. H_2O, as shown below. This structure shown the H – bonding in it. It explains the existence of $Cu(NH_3)_4$ SO_4H_2O like $Cu(H_2O)_4SO_4$. H_2O but no $CuSO_4$. $5NH_3$ is known because NH_3 molecules cannot form H – bonds so readily as H_2O molecules.

π–Hydrogen Bonding

In recent past, it has been proved that π–electrons of an olefinic or aromatic system act as base or proton acceptor in H–bonding interaction. These interactions are further of two types depending upon the presence of proton donor and proton centres in the molecules under consideration.

(a) **Inter molecular π-hydrogen bonding :** This type of bonding takes place between two or more molecules of the same or different compounds. Some examples having intermolecular π-hydrogen bonding are depicted.

(b) **Intra-molecular π-hydrogen bonding :** This type of bonding takes place between the groups within a molecule. Some examples having this type of bonding.

Characteristic of Hydrogen Bond

The various characteristics are.

(a) **Nature :** The hydrogen bond is formed due to the electrostatic forces which exist between the positive and negative ends (dipoles) of the same molecule or of different molecules of the same species.

In simple words, it is a case of van der Waal's forces. For example while the bond energy of van der Waal's attraction is approximately 4.2 kJ per mole, that of hydrogen bond has been found to vary between 14.6–42 KJ per mole. Therefore, it is wrong to consider the hydrogen bonding as van der Waal's attraction.

The bond energy of ordinary covalent bond is 3360–4200 KJ per mole. Thus, hydrogen bond is quite different from a covalent bond. Also, the formation of a hydrogen bond does not involve any sharing of electron pairs. Therefore, it is quite different from a covalent bond.

(b) **Electronegativity :** Hydrogen bond can form a bridge only between two atoms both of which are small in size, and highly electronegative. For example, hydrogen bonds are formed by elements like N, F and O.

Recent infra-red studies have shown that hydrogen bonding can also occur between any polar $X^{-}– H^{+}$ group and a highly polarizable atom or molecule. For example, the molecule of o–chlorophenol shows hydrogen bonding. Similarly, it may be mentioned that the stabilisation of chloral hydrate is to be explained by the formation of an intra–molecular hydrogen bond.

(c) **Bond length :** Some research workers have suggested that the distance between two hydrogen bonded atoms may be –2.45 A. But others have suggested that the bond length depends upon the electronegativity of two atoms forming hydrogen bond. Some values are given below.

Table 5

Compound	*Bond*	*Bond Length of Hydrogen Bond Å*
$(HF)_n$	F–H...F	2.55
HCOOH	O–H...F	2.67
$H_2O(s)$	O–H...O	2.76
$B(OH)_3$	O–H...O	2.74

There is evidence that for bond lengths of less than 2.44 Å the hydrogen bond is symmetrical, and that as the bond length increases, the arrangement X–H....X becomes less symmetrical. Electron diffraction analysis has shown that the three atoms in the boric acid are not collinear.

(d) **Position of proton :** Normally the proton does not lie in the middle of two electronegative atoms in the hydrogen bond. In hydrofluoric acid, the proton is situated midway between the two fluorine atoms whereas in ice, the proton is not at the mid point between two oxygen atoms.

(e) **Physico–chemical properties :** The effect of hydrogen bond formation on the properties of a compound may be seen as follows.

(i) **Melting points and boiling points :** It is found that HF, H_2O and NH_3 have high values of the melting and boiling points as compared to the hydrides of the other elements in the respective subgroup. This can be explained on the basis of association of their molecules due to hydrogen bonding which, therefore, requires more energy than usual to rupture the hydrogen bonds to set the molecules free to melt or vaporise.

(ii) **Dielectric constants :** Compounds which undergo polymerisation by hydrogen bond formation have abnormally high dielectric constants. A curve may be drawn showing the relation between the dielectric constants of certain liquids at 20°C and their dipole moments when the substances concerned are either in non-polar solvents or in gaseous state, so as to avoid dipole association.

From the curve it follows that if the polymer is linear as in HCN, the dipole moment will be three times that of the monomer. This is roughly in agreement with the observed value. But if polymerisation leads to *zig-zag* chain or a ring structure, the dipole moment of the polymer may be less than that of the monomer.

Theories of Hydrogen Bond

A hydrogen atom of configuration Is' can form only one covalent bond because only two electrons can be accommodated in the first quantum shell, and hence the possibility of hydrogen being covalent in hydrogen compounds is unlikely. A bicovalency for hydrogen (H--A....B) is not possible because it would involve the use of 2s or 2p orbitals which are of much high energies as to be totally useless for bonding. For this reason, we have to seek another explanation of the way in which hydrogen can link two atoms in the structure A–H....B. The various theories are :

(a) Electrostatic approach : In compounds of hydrogen with strongly electronegative and small sized atoms like, F, N, O, the shared electron pair between the two atoms lies much more nearer. This consequently results in the development of a fractional positive charge on the hydrogen atom. For example, where X is some electronegative atom.

When two H–X molecules approach each other, the $H^{+\delta}$ will exert a strong electrostatic force and so can attract $X^{-\delta}$ of other molecule. This attractive force between hydrogen and X leads to the formation of weak bond called hydrogen bridge or hydrogen bond.

Thus, the hydrogen bond may be regarded predominantly electrostatic in nature. However, it has also a small amount of covalent character and is mainly a case of simple dipole-dipole or ion-dipole interaction.

$$\overset{+\delta}{H} - \overset{-\delta}{X} \ldots \overset{+\delta}{H} - \overset{\delta}{X} \ldots \overset{+\delta}{H} - \overset{-\delta}{X}$$

Hydrogen bond is formed with such electronegative atoms which are having small cationic radii. Fluorine, oxygen and nitrogen are of this character.

Electrostatic approach has been supported by the following facts.

(i) The strongest hydrogen bond is formed between the most high electronegative elements.

(ii) The bonded atoms have the tendency to become linear due to the electrostatic repulsion.

The various weaknesses of this theory are as follows.

(i) This theory also offers no explanation for shortening of the distance in X–H.

(ii) This theory also offers no explanation for the symmetry of HF_2 – and O–H - - - O bonds in oxalic acid.

(iii) This theory does not explain the repulsion between the nuclei of X and H at short distances as found in the hydrogen bonded compounds. For example, the H–F distance in HF_2^- has been found to be only 20 pm more than that found in H–F.

(b) Valence bond treatment : Hydrogen bonding is not electrostatic. In terms of valence bond theory, it involves the following resonance structures.

$$\underset{\text{I}}{\overset{-\delta}{X} - \overset{+\delta}{H} \ldots \overset{\delta}{Y}} \rightleftarrows \underset{\text{II}}{H - X : Y} \rightleftarrows \underset{\text{III}}{X - H \longleftarrow H}$$

The structure (III) involves bicovalent hydrogen with X–H as normal covalent bond and H ← Y as long covalent bond. This structure does not make much contribution to the overall energy of the hydrogen bond. Structure (I) arises due to dipole interaction whereas structure (II) arises due to induced dipole interaction. As both these structures (I) and (II) make major contributions to the overall energy, the hydrogen bond may be considered to be electrostatic in nature. Few cases are only known in which three structures (I), (II) and (III) make equal contributions to the overall energy of the hydrogen bond. An example is HF_2^- ion in which all the three structures contribute equally to the bond.

$$F :^- H - F \rightleftarrows F - H : F^- \rightleftarrows F^- : H^+ : F^-$$

The resonance theory explains the various weaknesses (i) to (iii) of electrostatic approach. However, the resonance theory does not give correct value of the resonance energy. For example, the calculated resonance energy is 33 kJ $mole^{-1}$ while the experimental value is 25 kJ $mole^{-1}$

(c) Molecular orbital treatment : *Pimentel* has discussed hydrogen bonding in the HF_2^- ions using molecular orbital theory. He takes linear combinations of the hydrogen s orbital with fluorine p orbitals directed along the bond. Thus, if the p-orbitals of the two fluorine atoms are denoted by p_A and p_B the symmetrical F–H...F bond is described by

$$\Psi_1 \text{ (bonding)} = p_A + P_B + a_1 s$$

$$\Psi_2 \text{ (non – bonding)} = P_A + P_B$$

$$\Psi_3 \text{ (anti– bonding)} = P_A + P_B - a_3 s$$

where a_1 and a_3 are mixing co-efficients. The two electrons involved in bond formation are contained in a molecular orbital (Ψ_1) which spreads out axially on either sides of the hydrogen atom, thus forming two equivalent weak bonds, and there are also two electrons concentrated mainly on the fluorine atoms in the non-bonding orbital (Ψ_2). The molecular orbital diagram of HF_2^-.

Calculation of Hydrogen Bond Energy

According to *Allen* (1975), the hydrogen bond energy E of the hydrogen bond in H–A...B may be calculated by the following equation.

$$E = k\mu_{H-A}\ I/r$$

where $k = 2.44 \times 10^{18}\ C^{-1}$, μ_{H-A} = Bond moment of the H–A bond. I = Ionisation energy of the donor atom B and r = the distance H...B.

Conditions for Hydrogen Bonding

In case of inter-molecular hydrogen bonding, the most important condition is that the molecules must contain one hydrogen atom linked to one highly electronegative atom. In case of intramolecular hydrogen bonding the following conditions are favourable for hydrogen bonding.

(i) The molecule should contain two H–atoms linked to a highly electronegative atom and the other group should also contain a highly electronegative atom linked to a lesser electronegative atom.

(ii) The molecule should be planar.

(iii) The hydrogen bonding should lead to the formation of a five or six membered ring including the H–atom.

(iv) Interacting atoms must be arranged in such a way that there occurs minimum strain during the ring closure.

The condition (iv) is important only in the case of stereochemically rigid molecules. For example, the anti form of pyridine–2–carboxaldoxime undergoes intramolecular hydrogen bonding while the syn form of pyridine–2–carboxaldoxime does not undergo intramolecular hydrogen bonding because the lone pair on the pyridine N atom is not orienting towards the OH group of the oxime.

Properties Explained By Hydrogen Bonding

(a) State of H_2O and H_2S : Hydrogen bonding is responsible for the association of molecules in water for which reason it exists as a liquid. Since it is absent in H_2S, the latter exists as a gas.

(b) M.P. and B.P. of hydrides : The first member of the hydrides of group V A to VII A elements have abnormally high melting and boiling points relative to their molecular weights. Hydrogen bonding occurs in these compounds and therefore, more energy than usual must be supplied to the solids or liquids in order to rupture these bonds and thus set the molecules free to melt or vaporise. Thus, the relatively high melting and boiling points of NH_3. H_2O and HF can be explained on the basis of association of their molecules due to hydrogen bonding.

When the melting and boiling points of the hydrides of the elements of groups IV A, V A, VI A, and VII A, are plotted against the molecular weights of these hydrides, the plots shown in Fig. 1.16 are obtained.

(c) Some unique properties of water and ice :

(i) Structure of water : The oxygen atom of the water molecule can form, in addition to its two covalent bonds, two hydrogen bonds by association between its unbonded electron pairs and two hydrogen atoms from neighbouring molecules. From the figure, it is clear that the spatial arrangement of the two covalent bonds and two hydrogen bonds is tetrahedral, governed by the spatial arrangement of the bonded and unbonded electron pairs around oxygen. Due to the hydrogen bonding, H_2O exists as a liquid.

(ii) Structure of ice : In ice, the H_2O molecules are also held tetrahedrally respect to one another as shown in Fig. 1.19 but due to the comparatively lone O–H hydrogen bond in ice as compared to liquid water, ice has *open–cage* like structure. In this type of structure, the molecules are not packed very closely together whereas in liquid water they are packed closely together.

The crystal lattice structure of ice is shown in Fig. 1.18. The basic unit is hexagonal. Each oxygen atom is surrounded tetrahedrally by four hydrogen atoms, the two of which are covalently linked close to the oxygen atom at a distance of 1.0Å, and remaining two are associated by hydrogen bonds which are more remote at a distance of 1.75Å from the oxygen atom.

(iii) Density of ice is less than water : Ice has an open cage like structure in which lesser molecules are packed per c.c. When ice melts, the molecules come closer to one another. In another words, the molecules of water in the liquid state lie closer to one another than in ice. Consequently, the density of water in the liquid state is more than that in the solid state.

(iv) **Water has maximum density at 4°C but not at 0°C :** When ice is warmed, the liquid water is formed. But at 0°C, the liquid water still retains open cage like structure of ice.

The hydrogen bonds break up progressively with rise in temperature, the molecules pack closer together and the density increases from 0.999641 g/c.c. at 0°C to 0.99972 g/c.c. at 3.98°C. Only above this temperature the normal expansion of a liquid due to increasing thermal agitations overcomes this effect. Water then begins to shown the usual decrease in density, with increase in temperature. Thus, water has maximum density at 4°C, and not at 0°C.

(v) **Alcohols :** Alcohols contain the highly electronegative oxygen atom linked to hydrogen atom. Thus the association occurs in alcohol molecules as follows.

$$\begin{array}{ccc} R & R & R \\ | & | & | \\ H-O\ldots H- & O\ldots H- & O\ldots \end{array}$$

The fact that alcohols possess very high boiling points as compared to their corresponding sulphur analogues, thioalcohols (*mercaptans*) proves the existence of hydrogen bonding in alcohols.

	CH_3OH	C_2H_5OH	C_3H_7OH	C_4H_9OH
Boiling point	337.5 K	351 K	370 K	390 K
	CH_3SH	C_2H_5SH	C_3H_7SH	C_4H_9SH
Boiling point	278.8 K	310 K	340 K	370 K

(vi) **Carboxylic acids :** The carboxylic acids also contain the highly electronegative oxygen atom linked to the hydrogen atom. Hence, they also show association. In some of the carboxylic acids, they from dimers. This *dimer* is written as follows.

The formation of dimer is proved by the following facts.

(i) The electron diffraction studies confirm the presence of an eight–membered ring.

(ii) Vapour density measurements show the presence of doubly associated molecules.

It may be noted that in aqueous solution, the molecules of a carboxylic acid link up with water molecules rather than forming dimers.

(vii) **Dielectric constant and dipole moment :** The formation of a H bond A – H...B results in an increased polarity of the bond A – H and hence to a larger dielectric constant (permitivity) and greater dipole moment. We known that the permitivity) and greater dipole moment. We known that the permitivity is proportional to μ^2. Therefore, if two dipoles A–H, each of moment m associate to form a structure A–H...A–H of moment 2μ, the increase in permitivity is proportional to an increase from $2\mu^2$ to $(2\mu^2)$. This is considered to be the main reason for the abnormally large permitivity of water.

(vii) **Viscosity :** The increased organization of molecular structure in H bonded systems decreases the freedom of molecular motion, causes more "entanglements" between molecules and therefore, tends to increase the viscosity.

Detection Of Hydrogen Bond

The various method used for detecting hydrogen bonding in different compounds are.

1. **Infra-red spectroscopy :** When hydrogen bonding occurs between a proton donor group A–H and an acceptor group B, *i.e.*, a hydrogen bonded complex A–H–B is formed, several effects are usually found in the infra-red region.

   (a) The absorption bands due to the A–H stretching, virbrations (fundamentals and overtones) are shifted to lower frequencies. These shifts range from about 30 cm^{-1} to several hundred cm^{-1} or more. This effect is due to the weakening of the force constant for the A–H stretching mode, caused by the formation of the hydrogen bond.

   (b) New vibrational modes corresponding to H...B stretching and deformation, are found at low frequencies in the far–infrared region.

The fundamental frequency associated with OH stretching is 3600 cm^{-1}, if the OH group forms *intermolecular* hydrogen bonds, an additional band appears at 3300 cm^{-1}. Increase of temperature, or dilution of the solution weakens the intensity near 3300 cm^{-1}, and increases the intensity near 3600 cm^{-1}, thus, indicating that hydrogen bond formation is diminished by rise of temperature or fall of concentration. If an *intramolecular* hydrogen bond is formed as in salicylaldehyde, suppression of the 'free' OH group may be so complete that the band at 3600 cm^{-1} is absent, and only the band at 3300 cm^{-1} can be observed.

2. **Proton magnetic resonance :** This is very useful in the study of hydrogen bonding phenomena in liquids and solutions. The formation of a hydrogen bond A-H-B modifies the electron density around the proton of the A–H group and hence its shielding.

   In most cases proton magnetic resonance (PMR) absorption is displaced down field (lower magnetic field), causing a hydrogen bonding shift which is added to the existing chemical shift. Therefore, the hydrogen bonding shift is usually negative, with the exception of some cases involving aromatic molecules. The PMR method is useful in the study of weak inter-molecular bonds involving specific hydrogen atoms, *i.e.*, cases which exhibit small shifts and overlapping bands in their infra–red spectra.

3. **Electronic absorption spectroscopy :** Visible and U.V. spectroscopy in the visible region from 6000-2000Å involve excitation of electronic energy levels. If the chromophoric portion of a molecule is involved in hydrogen bonding, the electronic absorption or emission spectroscopy may provide experimental evidence about the hydrogen bonding.

   Experimentally it is observed that electronic transitions are shifted to the blue (hypsochromic shifts) when the chromophore is involved in hydrogen bonding as acceptor and shifted to the red (bathochromic shift) when serving as *donor*. Benzthiazoline-2-thione acts as a hydrogen bonding acceptor. It exhibits blue shifts in hydroxylic solvents due to the interactions – O–H ... $S = \overset{|}{C} -$ However, it shows red shifts in acceptor solvents-such as ketones which act as a hydrogen bonding donor –N–H... O = C–.

In those solvents where it can act as both hydrogen bond donor and acceptor (*e.g.*, self-association Benzthiazoline-2-thione in CCl_4) there is no shift in the UV absorption spectra.

The N–methyl derivative cannot be a hydrogen bond donor and shows only a blue shift in donor solvents.

4. **X-ray method :** In the usual method of X–ray diffraction, the position of the hydrogen atoms in a crystal are difficult to determine directly unless very accurate data are available and other atoms in the crystal are not too heavy. However, their positions may be inferred indirectly in the following way.

When the distances between certain pairs of atoms are unusually short, and the numbers of these short distances correspond to the number of hydrogen bonds required by the formula, then the short distances may be considered to indicate the positions of hydrogen bonds. For example, in the crystal of sodium hydrogen carbonate, two of the oxygens of each (planar) carbonate group are only 2.55Å from the nearest oxygens of adjacent groups, while the third is separated from its nearest neighbour. It is thus assumed that the short distances indicate the positions of the hydrogen atoms, and the resulting structure is an infinite chain ion in which discrete carbonate ions are linked through hydrogen bonds.

Importance of Hydrogen Bond

(I) Hydrogen bonds are important in fixing properties such as solubilities, melting points and boiling points and in determining the form and stability of crystalline structures.

(ii) Hydrogen bonds are important in biological systems. Proteins for example, contain both > CO and > NH group and hydrogen bonds can be formed to bridge the space between N and O. The structure and hence the properties of proteins depend upon the existence of hydrogen bonds.

(iii) Hydrogen bonding within a single molecule is one of the chief factors in determining the structure of important biological substances such as proteins.

BOND ENERGY OR STRENGTH

The process of bond formation involves release of energy, because the bonded atoms possess lesser energy than the individual atoms. It, therefore, follows that an equivalent amount of energy will have to be absorbed by the molecule for breaking the bond. The amount of energy required to break a mole of bond and separate the bonded atoms in a gaseous state is known as the bond energy.

Bond energy is a measure of the strength of the bond. • Bond energy is expressed in kJ or Kcal per mole. For a diatomic molecule like hydrogen, the bond energy is the same as that of bond dissociation energy because the breaking of one bond disrupts the whole molecule.

$$H_2(g) \rightarrow 2H(g) \text{ Bond energy} = 435.1 \text{ kJ mol}^{-1}$$

For polyatomic molecules, the bond energy is not equal to bond dissociation energy because in a polyatomic molecule, the energy necessary to rupture a given bond will depend to some extent on the nature of the remainder molecule. Therefore, the bond energy for a particular bond in polyatomic molecules will vary from compound to compound.

Calculations : One can calculate bond energies in the same way as that of bond lengths. For example, the energy of HI bond would be the sum of bond energy contribution of hydrogen and iodine atoms. Thus,

$$E_{H-1} = \frac{1}{2}E_{H-H} + \frac{1}{2}E_{I-1}$$

$$= \frac{435.4}{2} + \frac{150.6}{2} = 203.0 \text{ kJ mol}^{-1}$$

The calculated value of 293.3 kJ mol^{-1} is quite near to the experimental value of 297.0 kJ mol^{-1}. We can apply the above method of calculating bond energies to such molecules which have similar types of bonds.

The bond energies of some diatomic molecules are given in the following Table 6.

Table 6

Bond	*Bond Energy kj/mol*	*Bond*	*Bond Energy kj/mol*
H–H	435.4	C–C	347.8
O=O	497.9	C=C	606.7
N=N	945.6	C°C	803.3
F–F	154.8	C–O	334.7
Cl–Cl	242.7	C=O	694.5
I–Cl	209.2	C–S	272.0

The strength of the bond is indicated by the stability of the bond, *i.e.*, the bond energy. Thus, the strength of F–F bond is lower than that of Cl–Cl bond. Cosequently, fluorine is more reactive than chlorine.

Factors Affecting Bond Energy

(i) **Bond length :** Shorter the bond length, greater would be value of bond energy.

(ii) **Size of bonded atoms :** Shorter the size of bonds atoms, greater would be the value of their bond energy.

(iii) **Electronegativity :** Greater the electronegativity difference between bonded atoms, greater would be the value of their bond energy.

Importance : Its value has been regarded a measure of bond strength. Greater the bond energy, greater would be the strength of the bond.

BOND LENGTH

It is already mentioned that atoms involved in the formation of a bond cannot come closer to each other than a certain distance where the potential energy is minimum. The equilibrium distance between the centres of nuclei of two bonded atoms is known as bond length. It is expressed in Angstron (Å) units and can be determined with the help of X–ray diffraction, electron diffraction and spectroscopic methods. The actual bond length between two atoms depends on nature of atom, nature of bond and certain other factors. The average bond distances of some bonds are given below in A which is equal to 10^{-8} cm.

Table 7

Bond	*Bond length (Å)*	*Bond*	*Bond length (Å)*
H–H	0.74	H–Cl	1.27
Cl–Cl	1.99	C–H	1.09
C–C	1.54	C–Cl	1.77
C–C	1.33	O–H	0.96
C°C	1.20	C–O	1.42
C=O	1.24	C–N	1.09

One can understand the concept of bond length by considering the formation of H–H bond in a hydrogen molecule. In this figure, A_1 and A_2 represent the two nuclei while e_1 and e_2 represent the electrons of two hydrogen atoms.

When two hydrogen atoms approach each other, the following forces operate.

(i) The forces of attraction between electron e_1 and two nuclei ;

(ii) The forces of attraction between electron e_2 and two nuclei ;

(iii) The force of repulsion between electrons e_1 and e_2, and ;

(iv) The force of attraction between nuclei A_1 and A_2.

The forces of attraction bring the hydrogen atoms together while the forces of repulsion push them apart. But the forces of attraction are stronger than the forces of repulsion.

Therefore, a bond is said to be formed between the hydrogen atoms because *it is the force of attraction which results in the formation of chemical bond between the two hydrogen atoms.*

The distance between the two nuclei of two hydrogen atoms is the length of the H–H bond. Its value depends upon the balance between the forces of attraction and the forces of repulsion.

The X–ray diffraction method is generally used for measuring the bond lengths in the case of crystals while spectroscopy is mainly used for measuring the bond lengths in the case of gases.

Factors Affecting Bond Length

(i) **Hydrogen bonding :** The presence of H–bonding is able to shorten the bond length.

(ii) **Ionic character :** The bond length is found to increase with increase in ionic character.

(iii) **Bond multiplicity :** A multiple bond refers to that which is having a sigma (σ) bond and one or two pi (π) bonds. *e.g.*, C $\equiv$ C bond is made up of one sigma and two π bonds. Greater the number of π–bonds in a multiple bond, shorter would be the bond length.

Reason : A π-bond gets formed by the sidewise overlap of atomic orbital. When the number of π bonds gets increased the atomic orbitals (perpendicular to inter-nuclear axis) of bonded atoms come closer together for bringing about effective overlap. Thus, bond length gets decreased. For example, the bond length of C–C, C = C and C≡C have been 1.54 A, 1.34Å and 1.20Å respectively.

(iv) **s–character of orbitals :** The bond length has been found to decrease with increase in s–character of the orbitals forming a bond.

Importance of bond length : The magnitude of bond length provides the idea of single, double or triple bond between atoms.

3

Schematic Nomenclature of Coordination Compounds

INTRODUCTION

Coordination compounds are named according to the rules suggested by *International Union of Pure and Applied Chemistry*, IUPAC (1976). These rules are given below:

1. Order of naming cations and anions of an ionic complex compound

(i) If a complex compound iis ionic, *i.e.*, if a complex compound is composed of cations and anions, the name of the cation'is mentioned first and then the name of the anion is written. For Example:

(a) In naming $K_2[PtCl]$, the name of the cation, viz. K^+ is written first and then the name of the anion namely $[PtCl_6]^{2-}$ is mentioned.

(b) In naming $[Co(NH_3)_5Cl]Cl_2$, the name of the cation, $[Co(NH_3)_5Cl]^{2+}$ is written first and then the name of the anion, Cl^- is written.

(c) In $[Pt^{4+}(NH_3)_4Cl_2]^{2+}$ $[Pt^{2+}Cl_4]^{2-}$, the name of the cation, $[Pt^{2+}(NH_3)_4Cl_2]^{2+}$ is writen first and then the name of the anion, $[Pt^{2+}Cl_4]^{2-}$ is mentioned.

(ii) If the complex compound is non-ionic (neutral), the name of the complex compound is written as one word. Thus the name of $[Pt(NH_3)_4Cl_2]^0$ is written as one word.

2. How to name the species present in coordination sphere

In naming the species present in coordination sphere, the ligands are named first and than the name of the central metal atom is mentioned.

3. How to name the ligands?

(i) *How to name neutral ligands?*

(ii) *How to name negative (aniomc) ligands?* The names of negative ligands end in o.

(iii) *How to name positive ligands?* The names of positive ligands which are rare, end in ium.

(iv) *How to name free radicals?* Free radicals are given their own names. These are supposed to be neutral species, *i.e.*, they donot contain any charge on them.

(v) *How to indicate the number of ligands.*

(a) If a complex contains two or more simple ligands of the same type (*e.g.*, Cl^-, Br^-, NO_2^- $C_2O^{2-}{}_4$, NH_3, H_2O etc.), their number is indicated by putting prefixes di (for 2), tri (for 3), tetra (for 4), penta (for 5), hexa (for 6), septa (for 7), oct (for 8), nona (for 9), deca (for 10), undeca (for 11), dodeca (for 12) etc. before their names.

Table 6.1 : Examples of positive ligands and free radicals

Positive (cationk) ligands. The names of positive ligands end in ium, *e.g.*

NH_4^+	..... Ammonium	O_2^+	..... Oxygenyl
H_3O^+	..... Hydronium	NO_2^+	..... Nitronium
NO^+	..... Nitrosonium (also called nitrosyl, nitrolium or nitroso)	[H3N–NH2]+	..... Hydrazinium

Free radicals. Free radicals are given their own names *e.g.*,

CH_3	..... Ammonium	O_2^+	..... Oxygenyl
C_6H_5	..... Phenyl	CH_3CH	..... Acetonitrile

(b) If a complex contains two or more complex ligands or organic molecules which already have di, tri etc. in their names, their number is indicated by putting the prefixes bis (for 2), tris (for 3), tetrakis (for 4), pentakis (for 5), hexakis (for 6), hepta (for 7), octa (for 8), nona (for 9), deca (for 10), undeca (for 11), dodeca (for 12) etc. before their names. The name of the complex ligand or organic molecule is written in bracket. Examples of complex ligands are ethylene diamine (en), diethylamine, phosphorus trifluoride, acetylacetonato ion, benzoyl acctonate ion, dipyridyl, trichloro stannyl ion ($SnCl^-_3$), pyridine, methyl amine, triphynyl phosphine etc.

(vi) *How to name bridging ligands :* (a) If a complex has bridging ligands, the prefix μ (mi) is used before the name of each of the ligands. For example

$$[(NH_3)_5Co–NH_2–Co(NH_3)_4(H_2O)]Cl_5$$

is called pentanunine cobalt (in) u-amido tetrammine aquo cobalt (III) chloride or nonammine aquo (μ-amido) dicobalt (III) chloride.

(b) If a complex has two similar bridging ligands, \li-di is used for its name. For example:

$$\left[(H_2O)_4\ \overset{3+}{Fe} \begin{matrix} OH^- \\ \\ OH^- \end{matrix} \overset{3+}{Fe}\ (H_2O)_4\right](SO_4)_2$$

is called tetraquo iron (III) μ-dihydroxo tetraquo iron (III) sulphate or octaquo μ-di-hydroxo if-iron (III) sulphate.

(c) When the same ligand is present as a bridging ligand and as a non-bridging (terminal) ligand, the bridging ligand is named first. In case of the presence of three similar bridging ligands, μ-tri is used before their name. For example

$$\left[(CO)_3\ Fe \begin{matrix} CO \\ CO \\ CO \end{matrix} Fe\ (CO)_3\right]$$

is called m-tricarbonyl hexacarbonyl di-iron (0).

(d) If the bridging ligand bridges more than two metal atoms, m_3, m_4 etc. is used before the name of bridging ligand. m_3, m_4 etc. indicates the number of metal atoms between which the bridging ligand makes bridges, *e.g.* since in basic berylium acetate, $Be_4^{2+}O^{2-}(CH_3COO)_6$, *oxo* ($O^{2-}$) group briges four Be-atoms, in the name of this compound m_4 is used before the name of bridging ligand viz. O^{2-} group. Thus, the name of this compound is m_4-oxo-m-hexa acetato tetra beryllium (II).

(vii) *How to name ambidentate ligands :* See Ambidentate ligands and linkage isomerism discussed on page 18 (chapter 1).

(viii) *Order of naming ligands :* If the coordination sphere of a given complex compound contains various types of ligands, these ligands are named in alphabetical order. The prefixes di, tri etc. are not considered while determining this alphabetical order. For example $[Co(NH_3)_4Cl(NO_2)]^+$ ion is named as tetraaminechloronitrocobalt (III) ion.

The International Union of Pure and Applied Chemistry (IUPAC) publication *Nomenclature of Inorganic Chemistry* (1989), Blackwell scientific Publishers, contains the rules for the systematic naming of coordination compounds. The basic rules are summarized here.

1. The positive ion is named first followed by the negative ion.
2. When writing the name of a complex, the ligands are quoted in alphabetical order, regardless of their charge (followed by the metal).
3. When writing the formula of complexes, ligands are named before the metal. The coordinated groups are listed in the order: negative ligand, neutral ligands, positive ligands (and alphabetically according to the first symbol within each group).

(a) The names of negative ligands end in –o, for example :

F^-	fluoro	H^-	hydrido	HS^-	mercapto		
	Cl^-	chloro	OH^-	hydrxo	S^{2-} thio		
Br^-	bromo	O^{2-}	oxo	CN^-	cyano		
I^-	iodo	O_2^2	peroxo	NO^{2-}	nitro		

(b) Neutral groups have no special endings. Examples include NH_3 ammine, H_2O aqua, CO carbonyl and NO nitrosyl. The ligands N_2 and O_2 are called dinitrogen and dioxygen. Organic ligands are usually given their common names, for example phenyl, methyl, ethylenediamine, pyridine, triphenylphosphine.

(c) Positive groups end in-ium, *e.g.* NH_2–NH_2 hydrazinium.

4. Where there are several ligands of the same kind, we normally use the prefixes di, tri, tetra, penta and hexa to show the number of ligands of that type. An exception occurs when the name of the ligand includes a number, *e.g.* dipyridyl or ethylenediamine. To avoid confusion in such cases, bis, tris and tetrakis are used instead of di, tri and tetra and the name of the ligand is placed in brackets.

5. The oxidation state of the central metal is shown by a Roman numeral in brackets immediately following its name (*i.e.* no space, *e.g.* titanium (III)).

6. Complex positive ions and neutral molecules have no special ending but complex negative ions end in-ate.

7. If the complex contains two or more metal atoms, it is termed polynuclear. The bridging ligands which link the two metal atoms together are indicated by the prefix μ-.

   If there are two or more bridging groups of the same kind, this is indicated by di- μ, tri-μ-etc. Bridging groups are listed alphabetically with the other groups unless the symmetry of the molecule allows a simpler name.

   If a bridging group bridges more than two metal atoms, it is shown as μ_3, μ_4 , μ_5 or μ_6, to indicate how many atoms it is bonded to.

8. Sometimes a ligand may be attached through different atoms. Thus M–NO_2 is called nitro and M–ONO is called nitrito. Similarly the SCN group may bond M–SCN thiocyanato or M–NCS isothioc–yanato.

   These may be named systematically thiocyanato–S or thiocyanato–N to indicate which atom is bonded to the metal. This convention may be extended to other cases where the mode of linkage is ambiguous.

9. If any lattice components such as water or solvent of crystallization are present, these follow the name, and are preceded by the number of these groups in Arabic numerals.

These rules are illustrated by the following examples :

Complex anions	
$[Co(NH_3)_6]Cl_3$	Hexaamminecobalt(III) chloride
$[CoCl(NH3)5]^{2+}$	Pentaamminechlorocobalt(III) ion
$[CoSO_4(NH_3)_4]_4]NO_3$	Tetraamminesulphatocobalt(III) nitrate
$[Co(NO_2)_3(NH_3)_3]$	Triamminetrinitrocobalt(III)
$[CoCl.\ CN.\ NO_2.\ (NH_3)_3]$	Triamminechlorocyanonitrocobalt(ni)
$[Zn(NCS)_4]^{2+}$	Tetrathiocyanato–N–zinc(II)
$[Cd(SCN)_4]^{2+}$	Tetrathiocyanato–S–cadmium(II)
Complex cations	
$Li[AlH_4]$	Lithium tetrahydridoaluminate(III) *(lithium aluminium hydride)*
$Na_2[ZnCl_4]$	Sodium tetrachlorozincate(II)
$K_4[Fe(CN)_6]$	Potasssium hexacyanoferrate(II)
$K_3[Fe(CN)_5NO]$	Potassium pentacyanonitrosylferrate(II)
$k_2[CsCl_5N]$	Potassium pentachloronitridoosmate(VI)
$Na_3[Ag(S_2O_3)_2]$	Sodium bis(thiosulphato) argentate(I)
$K_2[Cr(CN)_2O_2(O_2)NH_3]$	Potassium amminedicyanodioxoperoxo chromate(VI)
Organic groups	
$[Pt(py)_4][PtCl_4]$	Tetrapyridineplatinum (II) tetrachloroplatinate (II)
$[Cr(en)_3]Cl_3$	*d* or l Tris (ethylenediamine) chromium (III) chloride
$[CuCl_2(CH_3NH_2)_2]$	Dichlorobis(dimethylamine)copper (II)
$Fe(C_5H_6)_2$	Bis(cyclopentadienyl)iron (II)
$[Cr(C_6H_6)_2]$	Bis(benzene)chromium (0)
Bridging groups	
$[(NH_3)_5Co.NH_2.\ Co(NH_3)_5](NO_3)_5$	μ– amidobis [pentaamminecobalt (III)] nitrate
$[(CO)_3Fe(CO)_3Fe(CO)_3]$	Tri-μ–carbonyl–bis(tricarbonyliron(0)) *(di–ironenneacarbonyt)*
$[Be_4O(CH_3COO)_6]$	Hexa–μ–acetato(O,O')–μ_4–oxotetraberyllium(II)*(basic beryllium acetate)*
Hydrates	
$AlK(SO_4)2.12H_2O$	Aluminium potassium sulphate 12 water

Stereochemical Descriptors

Different geometrical arrangements of the atoms attached to the central atom are possible for all coordination numbers greater than one.

The coordination polyhedron (or polygon in planar molecules) may be denoted in the name by an affix called the *polyhedral symbol.* This descriptor clearly distinguishes isomers differing in the geometries of their coordination polyhedra.

Given the same coordination polyhedron, diastereo-isomerism can arise when the ligands are not all alike as with the *cis* and *trans* isomers of tetraamminedichlorochromium (III), diamminedi–chloroplatinum (II), and bis (2–aminoethane–thiolato) nickel (II) (Examples 1–6).

Attempts to produce descriptors similar to *cis* and *trans* for stereochemically more complicated coordination entities have failed to achieve generality, and labels such as *fac* and *mer* are no longer recommended.

Nevertheless, a diastereoisomeric structure may be indicated for any polyhedron using a *configuration index* as an affix to the name or formula. Finally, the chiralities of enantiomeric structures can be indicated using *chirality symbols.*

Table 1 : Structural prefixes uses in inorganic nomenclature.

	These affixes are italized and separated from the rest of the name by hyphens
antiprismo	eight atoms bound into a rectangular antiprism
arachno	a boron structure intermediate between *nido-* and *hypho-* in degree of openness.
asym	asymmetrical
catena	a chain structure; often used to designate linear polymeric substances
cis	*two* groups occupying adjacent positions, not now generally recommended for precise nomenclature purposes
closo	a cage or closed structure, especially a boron skeleton that is a polyhedron having all faces triangular
cyclo	a ring structure. *Cyclo* here is used as a modifier indicating

	structure and hence is italicized. In organic nomenclature, 'cyclo' is considered to be part of the parent name since it changes the molecular formula, and therefore is not italicized.
dodecahedro	eight atoms bound into a dodecahedron with triangular faces
fac	three groups occupying the corners of the same face of an octahedron, not now generally recommended for precise nomenclature purposes
hexahedro	eight atoms bound into a hexahedron (*e.g.*, cube)
hexaprismo	twelve atoms bound into a hexagonal prism
hypho	an open strucuture, especially a boron skeleton, more closed than a *klado*–structure, but more open than an arachino-structure
icosahedro	twelve atoms bound into a triangular icosahedron
klado	a very open polyboron structure
mer	meridional; three groups occupying vertices of an octahedron in such a relationship that one is *cis* to the two others which are themselves *trans*, not now recommended for precise nomenclature purposes
nido	a nest–like structure, especially a boron skeleton that is almost closed
octahedro	six atoms bound into an octahedron
pentaprismo	ten atoms bound into a pentagonal prism
quadro	four atoms bound into a quadrangle (*e.g.*, square)
sym	symmetrical
tetrahedro	four atoms bound into a tetrahedron
trans	two groups directly across a central atom from each other, *i.e.*, in the polar positions on a sphere, not now generally recommended for precise nomenclature purposes
triangulo	three atoms bound into a triangle
triprismo	six atoms bound into a triangular prism
μ (*mu*)	signifies that a group so designated bridges two or more centers of coordination
λ (*lambda*)	signifies, with its superscript, the bonding number, *i.e.*, the sum of the number of skeletal bonds and the number of hydrogen atoms associated with an atom in a parent compound.

Polyhedral symbols : The polyhedral symbol indicates the geometrical arrangements of the coordinating atoms about the central atom.

It consists of one or more capital italic letters derived from common geometric terms which denote the idealized geometry of the ligands around the coordination center, and an arabic numeral that is the coordination number of the central atom. The polyhedral symbol is used as an affix, enclosed in parentheses, and separated from the name by a hyphen.

Configuration Index

Definition of index and assignment of priority numbers to ligating atoms. Having developed descriptors for the general geometry of coordination compounds, it becomes necessary to identify the individual coordination position.

The *configuration index* is e series of digits identifying the positions of the ligating atoms on the vertices of the coordination polyhedron. The individual configuration index has the property that it distinguishes between diastereoisomers. The digits of the configuration index are established by assigning an order of priority to the legating atoms as described. The procedure for assigning priority numbers to the ligating atoms of a mononuclear coordination system is based on the standard sequence rule developed for enantiomeric carbon compounds by Cahn, Ingold, and Prelog. The ligating atom with highest priority is assigned the priority number 1, the ligating atom with the next highest priority, 2, and so on.

Configuration Indexes For Particular Geometries

Square planar coordination systems (*SP* – 4). The configuration index in a single digit which is the priority number for the ligating atom *Irons* to the ligating atom of priority number 1.

Octahedral coordination systems (*OC* – 6). The configuration index consists of two digits. The first digit is the priority number of the ligating atom *trans* to the ligating atom of priority number 1. These two ligating atoms define the reference axis of the octahedron. The second digit of the configuration index is the priority number of the ligating atoms *trans* to the ligating atom with the lowest priority number (most preferred) in the plane that is perpendicular to the reference axis.

Table 2 : List of Polyhedral Symbols

Coordination polyhedron	*Coordination number*	*Polyhedral symbol*
linear	2	L–2
angular	2	A–2
trigonal plane	3	TP–3
trigonal pyramid	3	TPY–3
tetrahedron	4	T–4
square plane	4	SP–4
trigonal bipyramid	5	TBPY–5
square pyramid	5	SPY–5
octahedron	6	OC–6
trigonal prism	6	TPR–6
pentagonal bipyramid	7	PBPY–7
octahedron, face monocapped	7	OCF–7
trigonal prism, square face monocapped	7	TPRS–1
cube	8	CU–8
square antiprism	8	SAPR–8
trigonal prism, triangular face bicapped	8	TPRT–8
trigonal prism, square face bicapped	8	TPRS–8
trigonal prism, square face tricapped	9	TPRS–9
heptagonal bipyramid	9	HBPY–9
dodecahedron	8	DD–8
hexagonal bipyramid	8	HBPY–8
octahedron, trans-bicapped	8	OCT–8

Formulae and Names for Chelate Complexes Designation of the Ligating Atoms in a Polydentate Ligand

Donor atom symbol as the index : A polydentate ligand possesses more than one donor site, some or all or which may be involved in coordination. Thus, dithiooxalate anion conceivably may be attached through S or O, and these were distinguished by dithiooxalato-S, S′ and dithiooxalato-O, O′ respectively.

If the same element is involved in the different positions, the place in the chain or ring to which the central atom is attached is indicated by numerical superscripts.

Examples:

1.$(CH_3COCHCOCH_3)$ 2,4–pentanedionato–C_3

The kappa convention. As the complexity of the ligand name increases, a more general system is needed to indicate the points of ligation. In the nomenclature of polydentate chelate complexes, single ligating atom attachments of a polyatomic ligand to a coordination center are indicated by the italic element symbol preceded by a Greek kappa, k. Monodentate ambident ligands provide simple examples, although for these cases the kappa convention is not significantly more useful than the 'donor atom symbol' convention [above].

Nitrogen-bonded NCS [formerly isothiocyanato] is thiocyanato-kN and sulfur bonded NCS is thiocyanato-kS. Nitrogen-bonded nitrite [formerly nitro] is named nitrito-kN and oxygen-bonded nitrite is nitrito-kO, *i.e.,* $[O = N–O–Co(NH_3)_5]^{2+}$ is pentaamminenitrito-kO-cobalt (III) ion.

For polydentate ligands, a right superscript numeral is added to the symbol *k* in order to indicate the number of identically bound ligating atoms in the *flexidentate*[2] ligand. Any doubling prefixes, such as bis–, are presumed to operate on the locant index as well. Examples [1 and 2] use tridentate chelation by the linear tetraammine ligand, N,N′–bis(2–aminoethyl)1,2– ethanediamine to illustrate these rules.

Stereo chemical descriptors for chelated complexes. Stereo-chemical descriptors can be provided for compounds containing chelated ligands but they involve considerations beyond those described above. The polyhedral symbol is determined as in the case of monodentate ligand derivatives. Also, the priority numbers are assigned to ligating atoms as for monodentate ligands.

However, a general treatment for the assignment of the configuration index requires the use of priming conventions in order to provide a completely systematic treatment. Thus, for a particular diastereoisomer of $[Co(NH_3)_2(NH_2CH_2CH_2\ NH_2)Br_2]^+$, the polyhedral symbol is OC–6 and the ligating atom priority number are as shown below.

For this case, there are no additional complications, and the configuration index is assigned in the usual way as OC–6–32. The classic case of diastereoisomerism that arises among chelate ligand derivatives is the tris(didentate) complexes in which the two donor atoms of the identical ligands are different. Glycinate, $NH_2\ CH_2\ CO_2^-$

and 2–aminoethanethiolate, $NH_2CH_2CH_2S^-$ illustrate this. For complexes of either ligand, the facial and meridional labels described previously could be applied, but the more systematic configuration indexes are *OC*–6–22 and *OC*–6–21.

Priming convention. The configuration index is especially useful for bis(tridentate) complexes and for more complicated cases. Bis (tridentate) complexes exist in three diastereoisomeric forms which serve to illustrate the utility of a priming convention. These isomers are represented below, along with their site symmetry symbols and configuration indexes. For examples 1, 2 and 3, the two ligands are identical and the ligating-atom priority numbers are indicated.

Chirality Symbols

Symbols *R and S*. There are two established and well-used systems for chirality symbols and these differ in fundamental ways. The first, the convention for chiral carbon atoms is equally appropriate to metal complexes and is most often used in conjunction with ligand chirality. However, it can be applied to metal centers and has been useful for pseudotetrahedral organometallic complexes when, for example, cyclopentadienyl ligands are treated as if they were monodentate ligands of high priority.

Skew-line convention for octahedral complexes. The second is the skew line convention and applies to octahedral complexes. Tris(didentate)complexes constitute a general family of structures for which a useful unambiguous convention has been developed based on the orientation of skew lines which define a helix. Examples 1 and 2 represent the *delta* and *lambda* forms of a complex, such as $[Co(NH_2CH_2CH_2NH_2)^{3+}$.

Chirality symbols based on the priority sequence. The procedure is applied as for tetrahedral, but modified because there is a unique principal axis. The structure is oriented so that the viewer looks down the principal axis, with the ligand having the higher priority closer to the viewer. Using this orientation, the priority sequence of ligating atoms in the [horizontal] plane is examined.

If (he sequence proceeds in a clockwise fashion, the chirality symbol *C* is assigned. Conversely, if the sequence is anticlockwise, the symbol *A* is assigned.

Polynuclear Complexes

Polynuclear inorganic compounds exist in a bewildering array of structural types, such as ionic solids, molecular polymers, extended assemblies of oxoanions both of metals and nonmetals, nonmetal chains and rings, bridged metal complexes, and homo-and hetero-nuclear clusters. This section treats primarily the nomenclature of bridged metal complexes and homo-and hetero-nuclear clusters.

Bridging ligands, as far as they can be specified, are indicated by the Greek letter μ appearing before the ligand name and separated by a hyphen. The whole term, *e.g.*, μ-chloro, is separated from the rest of the name by hyphens, as in ammine-μ-chloro-chloro, etc., or by parentheses if more complex ligands are involved.

The bridging index, the number of coordination centers connected by a bridging ligand, is indicated by a right subscript, μ_n, where $n > 2$. The bridging index 2 is not normally indicated. Bridging ligands are listed in alphabetic order along with the other ligands, but a bridging ligand is cited before a corresponding nonbridging ligand, as with di–μ–chloro-tetrachloro.

Examples

1. $[CoCu_2Sn(CH)_3)]\mu-(C_2H_3O_2)_2(C_5H_5)]$

   bis(μ-acetato) (cyclopentadienyl)(methyl)–cobaltdicoppertin

2. $[\{Cr(NH_3)_5\}_2(\mu\text{–}OH)]Cl_5$

   μ-hydroxo-bis(pentaamminechromium)(5+)– pentachloride

3. $[[PtCl\{P(C_6H_5)_3\}]_2(\mu-Cl)_2]$ di-μ-chlorobis[chloro (triphenyphosphine)–platinum]

4. $[\{Fe(NO2)2\}2\{\mu\text{–}P(C6H_5)_2)_2]$

   bis (μ–diphenylphosphido)bis(dinitrosyliron)

Examples

1. $[Br_4ReReBr_4]^{2-}$ bis(tetrabromorhenate)(Re–Re)(2–)
2. $[Mn_2(CO)_{10}]$ bis(pentacarbonylmanganese)(Mn–Mn) or decacarbonyldimanganese (Mn–Mn).

SPECTROCHEMICAL SERIES

One of the factors which affects the degree of splitting of d–orbitals (10D or Δ) is the *nature of the ligands.* The value of 10Dq can be equated to the energy observed as the maximum in the absorption spectra of complexes as explained earlier for $[Ti(H_2O)_6]^{3+}$. ON this basis, ligands may be arranged in order of their field strength, *i.e.,* the order of energy of the first absorption bands of the complex.

$$I^- < Br^- < S^{2-} < SCN^- < Cr^- < NO_3^- < F^- < OH^- < Ox^{2-} <$$

$$H_2O < NCS^- < NH_3 < en < dipy < phen < NO_2^- < CN^- < CO$$

This arrangement of ligands in the order of *increasing field strength* is called Fajans-Tschida *Spectrochemical series* because it is derived from spectral studies of the complexes. The ligands on the right side of the series provide the strongest field (high 10Dq) are the most effective in forcing the cations into low spin states whereas the ligands on the left, provide weak field (low 10Dq) add usually give high spin complexes.

For example we may consider the formation of two octahedral complexes of Fe^{3+} $[Fe(H_2O)_6]^{3+}$ and $[Fe(CN)_6]^{3-}$ where Fe^{3+} has d^5 configuration. Since H_2O is a weaker ligand than CN^- ion, the value of splitting energy (10 Dq) is smaller in $[Fe(H_2O)_6]^{3+}$ than in $[Fe(CN)_6]$. Hence, the five d electrons of Fe^{3+} will occupy all the t_{2g}, and e_g levels without pairing, resulting in a high spin complex with 5 unpaired electrons.

Thus $[Fe(CN)_6]^{3-}$ is a highly paramagnetic complex. In case of $[Fe(CN)_6]^{3-}$ the value of 10 Dq is sufficiently high to force pairing of t_{2g} electrons. Thus in $[Fe(CN)_6]^{3-}$ there are two pairs of electrons in two t_{2g} orbitals and single electron in one t_{2g} orbital. Hence it is also paramagnetic in nature but it is less paramagnetic as compared to $[Fe(H_2O)_6]^{3+}$.

The effect of the order of increasing ligand field strength in spectrochemical series, will be clear from the absorption spectra of some complexes of Co^{3+} and Cr^{3+}.

As there is a steady progress in the frequency of the absorption maxima as the ligand changes from $H_2O \rightarrow NH_3 \rightarrow CN$.

In can see that there is a steady, progress in frequency of the absorption maxima as the ligands change with the linking atoms progressing $Cl \rightarrow S \rightarrow O \rightarrow N \rightarrow C$ corresponding to a progressive increase in the value of 10 Dq. Though, spectrochemical series allows us to rationalise differences in spectra, it creates many serious problems which could not be accounted for, on the basis of crystal field theory. If the ligands arc considered as point charges or dipoles (as assumed by CFT) then anionic ligands should bring out more crystal field splitting than the neutral molecules.

According to CFT, among negative ions F^- is expected to be a stronger ligand than CN^- and among neutral molecules H_2O is expected to be stronger than NH_3 (H_2O has higher dipole moment than NH_3). But in the spectrochemical series, their positions are just the reverse. The position of these ligands can be explained only when π bonding is assumed between the central metal and the ligand (*Molecular orbital theory*).

Since CFT completely neglects π bonding, it cannot account for the position of all the ligands in the spectrochemical series. This failure to interpret the ligand field strength remains as one of the major drawbacks of the crystal field theory.

APPLICATIONS OF CRYSTAL FIELD THEORY

The various applications are as follows :

1. Colour of Transition Metal Complexes : Most of the complexes of transition metals give absorption bands in the visible region. It means that they are coloured. According to the crystal field theory, the colour is due to the *d-d* transitions between the *eg* and t_{2g} electrons. The energy difference between the e_g and t_2g electrons is sc small that even absorption of low radiation in the visible light may cause excitation from a lower to a higher *d* level.

As a result of the absorption of some selected wave lengths of visible light, the complexes appear coloured. For example, the complex $[Ti(H_2O)_6]^+$ absorbs green and yellow light from the white light, it means that the complex will have blue and red colour *i.e.*, purple colour.

The colour intensity of a transition metal complex depends upon the magnitude of Δ. For example, the value of Δ for $[Co(NH_3)_6]^{2+}$ is very large. It means that it requires relatively large excitation energy which it gets from the blue portions of the visible light. Its aqueous solution, therefore, appears red (white = blue + red). The magnitude of Δ depends upon electronic configuration, size and oxidation state of metal ion, nature of ligand and geometry of the complex.

2. **Magnetic Properties of Complexes:** The crystal field theory is successful in calculating the number of unpaired electrons in high spin and low spin complexes. With the help of number of unpaired electrons, one can find the value of 'spin only moment' by applying the formula:

$$\mu_s = \sqrt{[n(n+2)]} \text{ B.M.}$$

From a knowledge of unpaired electrons and spin only moment, it is possible to know two important aspects of complexes.

(i) The valence state of the metal ion in an octahedral complex can be calculated.

(ii) If there are more than 3d electrons, the nature of bonding in a complex may be known.

These two points may be understood by undergoing the following examples.

(i) The magnetic moment of $Hg[Co(CNS)_4]$ is 4.5 B. M. From this value, it is evident that cobalt is in the bivalent state ($Co^{2+} \rightarrow d^7$) and the bonding belongs to spin-free type corresponding to n = 3.

(ii) The magnetic moment of $[Fe\ (diph)_3]\ (ClO_4)_3$ is 2–2 B.M. From this value, it is evident that iron is in the trivalent state ($Fe^{3+} \rightarrow d^5$) and the bonding belongs to spin paired type which corresponds to n = 1.

The values of magnetic moments also help to decide whether a given octahedral complex is a high-spin or a low-spin complex. If the value of μ_{exp} value for the complex is less than the μ, value for the metal ion, the complex belongs to low-spin type otherwise it belongs to high-spin type.

3. **Relative liabilities of Complexes :** The crystal field theory gives useful information about relative liabilities of certain complexes and also provides information about the mechanism of their substitution reactions. For example,

(i) From the crystal field stabilisation energy calculations of strong field octahedral complexes, it is evident that the speed of d^0, d^1 and d^2 complexes will be greater than the corresponding d^3, d^4, d^5 and d^6 complexes. This speed is independent of $S_N{}^1$ and $S_N{}^2$ mechanism.

(ii) CFSE calculations reveal a decrease in the order of reactivity of a, d^5, d^4, d^3, d^6 complexes.

(iii) When CFSE calculations are made for weak-field octahedral complexes, it follows front these values that complexes of d^n configuration except d^3 and d^8 should lead to fast reactions irrespective of $S_N{}^1$ and $S_N{}^2$ mechanism.

(iv) CFSE calculations also predicted that d^8 complex (Ni complex) should be inert.

All the above conclusions are in full agreement with the experimental evidences.

4. Stabilisation of Oxidation States : The crystal field theory explains why certain oxidation states are preferentially stabilised by coordination to specific ligands. For example, water stabilises Co(II) whereas ammonia stabilises cobalt (III). This is evident from the following CFSE data :

5. Stereochemistry of Complexes : The crystal field theory explains that copper (II) forms square planar complexes than tetrahedral or octahedral complexes The reason for this is that copper (II) possesses a much higher CFSE. (12.28Dq) in a square planar configuration than octahedral (CFSE = 6Dq) or tetrahedral configuration (CFSE= – 2.56Dq).

The CFSE also explains that the square planar complexes of 4d and 5d metal ions form more readily than *3d* ions.

Table 3

	CFSE (-Dq) Weak field (H_2O)	*CFSE (-Dq) Strong field (NH_3)*
$Co^{2+}(d^7)$	7	18
Co^{3+} (d^6)	4	24

6. Distorted Octahedral Complexes : In an octahedral complex, six ligand molecules are arranged around the central metal ion. When the bond lengths are the same for all six ligands, the octahahedral is said to have a regular shape.

If the bond lengths are not same in all six ligands, the shape of octahedral complex is distorted. This change in shape is known as *distortion* and the complex is called *distorted octahedral complex.*

In 1937 Jahn–Teller theorem was put forward to explain why certain six coordinated complexes possess distorted octahedral geometry.

According to this theorem,

(i) An octahedral complex is said to have a regular shape if the d orbitals (both t_{2g} and e_g sets) are occupied symmetrically.

(ii) When d-orbitals of the central metal ion of an octahedral complex possess t_{2g} orbitals as unsymmetrical orbitals, there occurs slight distortion from the regular octahedron. This distortion is due to unevenly filling in an octahedral complex because inese orbitals do not point towards the ligands. The slight distortion may occur in such cases when the central metal ion contains 1, 2, 4 or 5 electrons.

(iii) Strong distortion occurs when the e_g orbitals of an octahedral complex are unsymmetrically filled because these orbitals do not point towards the ligands. Due to this strong distortion the octahedral shapes of complexes may change to tetragonal and even to square planar complexes. The high-spin complexes of d^4 and d^9 and low spin complexes of d^7, d^8 and d^9 lead to strong distortion.

The above three postulates (i), (ii) and (iii) which describe the effect of unsymmetrical t_{2g} or e_g orbitals on the shape of an octahedral complex is known as Jahn–Teller effect.

Explanation : The crystal field theory explains Jahn–Teller effect. This can be illustrated by considering the distortion produced by the presence of e_g orbitals in a complex of Cu^{2+} ion (d^9 ion). This ion possesses the configuration $t_{2g}^6\ e_g^3$ in both the fields. Thus, there are two possible configurations L :

$$\text{I} \longrightarrow t_{2g}^6\ (d_z^2)^2\ (dx^2 - y^2)^1$$

$$\text{II} \longrightarrow t_{2g}^6\ (d_z^2)^2\ (dx^2 - y^2)^1$$

As both the configurations (t_{2g}) orbitals) are completely field, it means that asymmetry is due to the incomplete filling of e_g orbitals. Thus, distortion is mainly due to the repulsion of ligands by the electrons occupying e_g orbitals.

In configuration I the doubly occupied d_z^2 orbital offers greater shielding to the nuclear charge of the copper than the half filled $dx^2 - y^2$ orbitals. The result of this is that :

(i) The ligands along z-axis which realise lower effective nuclear charge tend to move away from Cu^{2+} nucleus, and

(ii) The ligands along xy plane which experience a higher effective nuclear charge tend to move closer.

From the above two observations (i) and (ii), it is evident that the complex of copper should have four short and two long metal ligand bonds.

If the electronic configuration II is accepted, it is expected that there will be a distortion but in the opposite direction in which the configuration I is predicting. According to conlguration II, the net result is :

(i) The ligands along z-axis tend to move closer to the nucleus and

(ii) The ligands along xy plane would move away.

From the above two results (i) and (ii), it follows that the complex of copper should have four long and two short metal-ligands bonds.

Theoretically, there is no explanation for deciding which of the two possible distortions is more likely. However, experimental results have revealed that distortions are more common in those octahedral complexes in which there are four short and two long bonds. For example,

(i) In the cupric chloride crystal each Cu^{2+} ion is surrounded by six Cl^- ions. Four of them are lying at a distance of 2.30 Å and the other two are at 2.95 Å away.

(ii) In cupric fluoride crystal F^- ions are lying at a distance of 1.93Å and the remaining two F^- ion are at 2.27 Å away.

We will now consider the effect of distortion due to Jahn–Teller effect on the energy of Cu^{2+} ion (d^9).

In splitting of the more stable octahedral distortion corresponding to configuration I $t_{2g}6$ $(dz)^2$ $(dx^2 - y^2)^1$ is being considered. δ_1 and δ_2 represent the splitting of the e_g^{3-} and t_{2g}^6 – levels respectively. Both δ_1 and δ_2 are much smaller as compared with Δ_0 and also, δ_2 is much smaller than δ_1 *i.e.*,

$$\Delta_0 >>> \delta_1 > \delta_2$$

Under these conditions both t_{2g} and e_g orbitals are split. For t_{2g} electrons there is no net energy exchange because four electrons (d_{yx}

and d_{zx}) are stabilised by $4 \times (-1/3\ \delta_2) = -4/3\ \delta_2$ and the remaining two electrons (d_{zy}) get stabilised by $(2 \times 1/3\ \delta_2 + 2/3\ \delta_2) = 4/3\ \delta_2$. Thus,

Net energy for t_{2g} electrons = energy gain + energy loss

or Net energy gain for t_{2g} electrons $= -4/3\ \delta_2 + (4/3\ \delta_2) = 0$.

In the splitting at e_g levels, two electrons are stabilised by $2 \times \left(-\frac{1}{2}\delta_1\right)\delta_1$ while one electron is destabilised by $1 \times \left(+\frac{1}{2}\delta_1\right) = \frac{1}{2}\delta_1$, *i.e.*,

Net energy gain for e_g electrons = energy gain + energy loss

$$= +1/2\ \delta_1 - \delta_1 = -\delta_1/2.$$

For the above simple calculations, the net energy change is $\delta_1/2$ which provides the driving force of distortion. This energy is called *John-Teller stabilisation energy.*

Failures of C.F.T.

(i) In CFT model metal-ligand bonding is taken to be purely electrostatic which is quite unrealistic. It provides no account of partial covalent character of the metal- ligand bonds.

(ii) The theory cannot explain for the relative strengths of ligands. For example, it provides no account why OH^- has a smaller splitting than H_2O.

(iii) CFT considers only the metal ion d-orbitals without giving any consideration to other metal orbitals such as s, p_x p_y and p_z and ligand π-orbitals.

(iv) CFT is unable to consider the possibility of the existence of π-bonding in complexes despite of its frequent existence in complexes of metals in low or high oxidation states.

(v) Crystal field theory keeps the total emphasis on the metal orbitals without consideration to the ligand orbitals. Therefore, all properties dependent upon the ligand orbitals and their interaction with metal orbitals are not at all explained.

OCTAHEDRAL COMPLEXES

(a) Formation of Group or Composite Ligand Orbitals : It is clear that each of the metal orbitals would be in a position to overlap simultaneously with several ligand orbitals which may in turn belong to different ligands. But the same is not true for ligand orbitals. Before overlapping, the six ligand orbitals combine together linearly to form six group σ-orbitals that would be capable of overlapping with the central metal ion.

The determination of such linear combination of ligand o-orbitals is made by inspection method which is a straight forward and commonsense method. The linear combinations of ligand orbitals are called composite ligand orbitals.

The positions of six ligand σ–orbitals in an octahedral complex. The formation of group ligand σ–orbitals is shown pictorially.

(i) If the +ve lobes of a orbitals of all the six ligands in an octahedral complex are pointing towards the central metal ion then the composite orbital, signified by $(\sigma_x + \sigma_{-x} + \sigma_y + \sigma_{-y} + \sigma_{-z})$ is capable of overlapping with the s orbitals of the metal cation. This composite orbital is designated as Σ_a and its value when normalised is as follows :

$$\Sigma_a = \frac{1}{\sqrt{6}} (\sigma_x + \sigma_{-x} + \sigma_y - \sigma_{-y} + \sigma_z + \sigma_{-z})$$

(ii) Similar to above, for overlapping with p orbitals, the composite orbitals can be find out. For px, it would be $\sigma_x - \sigma_{-x}$. Similarly for p_y and p_z these would be $\sigma_y - \sigma_{-y}$ and $\sigma_z - \sigma_{-z}$. The composite orbitals p_x, p_y and p_z are designated as Σ_x, Σ_y, and Σ_z, respectively. Their values, when normalised, are follows.

$$\Sigma_x = \frac{1}{\sqrt{2}} (\sigma_x - \sigma_{-x})$$

$$\Sigma_y = \frac{1}{\sqrt{2}} (\sigma_y - \sigma_{-y})$$

$$\Sigma_z = \frac{1}{\sqrt{2}} (\sigma_z - \sigma_{-z})$$

(iii) For $dx^2 - y^2$, the composite orbital would be $(\sigma_x - \sigma_{-y} - \sigma_y - \sigma_{-y})$. Its normalised value is

$$\Sigma_x{}^2 - y^2 = \frac{1}{2} (\sigma_x + \sigma_{-x} - \sigma_y - \sigma_{-y})$$

(iv) For $d_z{}^2$, the composite orbitals would be $(2s_z + 2s_{-z} - r_x - s_{-x}\ s_y - s_{-y})$ Its normalised value is

$$\Sigma_{z^2} = \frac{1}{2\sqrt{3}} (2\sigma_z + 2\sigma_{-z} - \sigma_x - \sigma_{-x} - \sigma_x - \sigma_y - \sigma_{-y})$$

In the above equation the value of 2 before σ_z and σ_{-z} is the value of C ($C\psi_M = C_1\psi_1 + C_2\psi_2$).

The value of C decides the magnitude of contribution of that atomic orbital towards the formation of molecular orbital.

(b) Formation of Molecular Orbitals : In the next step, six atomic orbitals of the central metal ion.

$4s$, $4p_x$, $4p_y$, $4p_z$, $3d_z^2$ and $3d_x^2$ overlap with six groups of composite ligand orbitals formed as in step (b)) forming six sigma bonding and six sigma antibonding molecular orbitals. The bonding molecular orbitals are designated as σ^b – MO's whereas the antibonding molecular orbitals are designated as σ^* – MO's. Thus,

(i) 4_S *and* Σ_a *on overlapping give* σ_s^b *and* σ_s^* *MO's.*

(ii) $4px$ *and* Σ_x *on overlapping give* σ_x^b *and* σ_s^* *MO's.*

(iii) $4py$ *and* Σ_y *on overlapping give* σ_x^b *and* σ_y^* *MO's,*

(iv) $4p_z$ *and* Σ_z *on overlapping give* Σ_y^b *and* σ_z^* *MO's.*

(v) $3dx^2 - y^2$ *and* $\Sigma x^2 - y^2$ *on overlapping give* $\sigma^b\ x^2 - y^2$ *and* σ^* $x^2 - y^2$ *MO's.*

(vi) $3d^2$, *and* Σ_z^2 *on overlapping 'give* $\sigma^b z^2$ *and* $\sigma^* z^2$ *molecular orbitals.*

The remaining three orbitals on central metal ion, *i.e.*, $3d_{xy}$, $3d_{yz}$ and $3d_{zx}$ are non-bonding orbitals. Thus, there are is molecular orbitals available for electron filling. All the fifteen molecular orbitals. The number of electrons which each group of orbitals can hold is given in brackets.

(c) Filling of Molecular orbitals : Filling of the molecular orbitals occurs according to the Aufbau's principle. With d^n complexes, the t_{2g} and e_g* orbitals will fill up in just the same way as in crystal field theory. Let us illustrate this filling by considering a d^3 octahedral complex like $[Cr(NH_3J)_6]^{3+}$. In this complex, there is a total of 15 electrons (twelve from six metal ligand orbitals overlapping and three from metal *d* orbitals). These electrons are to be accommodated. Out of 15 electrons, twelve electrons are accommodated in the six bonding MO, *i.e.*, A_{1g} T_{1g} and E_g

whereas the remaining three go to the triply degenerate *i.e.,* non-bonding T_{2g} orbitals in the unpaired state. The energy difference between T_{2g} and E_g^* degenerate sets of orbitals is referred to as Δ. This is the same as Δ_0 of the complexes described in crystal field theory.

The only difference is that in ligand field theory, the value of Δ arises due to the different amount of electrostatic repulsion suffered by the various metal *d* electrons from the ligands whereas in molecular orbitals theory, Δ arises due to the metal-ligand covalent bonding which produces one set of non-bonding orbital T_{2g} and another of high energy antibonding orbitals (E_g^*) Thus, in molecular orbital theory, the concept of high and low spin complexes remains same as in crystal field theory. The magnitude of Δ decides the choice between the high and low-spin complexes. For example, in the hexafluoro complex of Co^{3+} the value of Δ is small and the distribution of electrons between and T_{2g} and E_g^* occurs as T_{2g}^4, E_g^{*2}. This makes it a high spin complex.

This distribution also explains why the Co–F bonds in the complex are not very strong. The reason for this is that the presence of two electrons in the antibonding orbitals reduces the strength of Co–F bonds. It is also important to remember that high spin complexes contain electrons in the antibonding orbitals This also explains why such complexes are relatively less stable.

(d) π-Bonding in Octahedral Complexes : The metal orbitals used for π bonding are as follows

$$d_{xy}, d_{yz}, d_{zx} - t_{2g} \text{ or } T_{2g}$$

and

$$p_x, p_y, p_z - t_{1u} \text{ or } T_{1u}$$

From the above it follows that t_{1u} orbitals are involved in o as well as π-overlaps whereas t_{2g} involve in π-overlaps. The ligand orbitals which take part in the π-overlaps are generally *p*π or *d*π orbitals. In the positions of various *p*π orbitals of the ligands are shown. In these the arrow indicates the direction of positive lobes of these orbitals. Each ligand will have two such *p*π orbitals at right angles to each other.

The composite ligand orbitals for these can be evolved as before. For example, the composite ligand orbitals of the *px (t1u)* orbitals of the metal would be $\pi_{3x} + \pi_{4x} + \pi_{5x} + \pi_{6x}$. For the d_{xz} t_{2g} orbital of the metal, the composite ligand orbital would be $(\pi_{1z} + \pi_{2z} + \pi_{5z} + \pi_{6x})$. Such types of π-orbitals are present in the ligand like oxides and fluorides. These orbitals have lower energy than the metal-π orbitals.

In π-overlaps, eighteen molecular orbitals (six metals orbitals and twelve ligand orbitals, 2π-orbitals on each ligand) are formed. Out of these six are bonding and [T_{2g} (3) and $T_1\pi$ (3) and six are non-bonding [$T_{2g}\pi^*$ (3) and $T_{2g}\pi$ (3)]. It is important to remember that T_{1g}^* and T_{1g}^* have the same ligand combination as T_{1g} and $T_{2g}\pi$, but with every other sign in the linear combination reversed. Similarly, $T_{1g}\pi^*$ and $T_{1u}\pi^*$ have the same combination as $T_{2g}\pi$ and T_{1u} but with every sign reversed.

The various π-bonding MO formed and the proper metal and ligand orbital combination required for them in an octahedral complex are given in Table 4.

Table 4

MO (Complex)	*Metal orbitals*	*Ligand orbital combinations*
T_{1u} (π)	p_x (t_{1u})	$(\pi_{3x} + \pi_{4x} + \pi_{5x} + \pi_{6x})$
	p_y (t_{1u})	$(\pi_{1y} + \pi_{2y} + \pi_{5y} + \pi_{6y})$
	p_z	$(t_{1u})(\pi_{1z} + \pi_{2z} + \pi_{3z} + \pi_{4z})$
T_{1u} (π)	d_{xy}	$(\pi_{1y} + \pi_{2y} + \pi_{3x} + \pi_{4x})$
	d_{yz}	$(\pi_{3z} + \pi_{1z} + \pi_{5u} + \pi_{6y})$
	d_{zx}	$(\pi_{1z} + \pi_{2z} + \pi_{5z} + \pi_{6x})$

The magnitude of Δ increases according to the nature of the ligands in the following order Strong π donor < weak π–donor < negligible

π-interaction < weak π–acceptor < strong π-acceptor.

This confirms the order of spectrochemical series.

(e) Central Metal Ion Orbitals. In an octahedral complex the metal ion is surrounded by six ligands placed along the $+x$, $-x$, $+y$, $-y$, $+z$ and $-z$ axes. We know that there are 9 atomic orbitals on the central metal atom if it belongs to 3*d* series. These are one outer 4s, three outer *p* ($4p_x$, $4p_y$ and $4p_z$) and five inner 3*d* ($3d_x^2 - y^2$, $3d_z^2$ $3d_{xy}$, $3d_{yz}$, $3d_{zx}$) atomic orbitals. These orbitals can be divided into two groups.

(i) The first group includes 4s, $3p_x$, $4p_y$, $4p_z$ $3d_z^2$ and $3d_x - y^2$ atomic orbitals. These orbitals have their lobes lying along the axes. The orbitals by overlapping with ligand orbitals form six metal-ligand σ bonds or molecular orbitals.

(ii) The remaining three atomic orbitals of central metal atom do not participate in σ-bonding and hence remain non-bonding. These can, however, take part in π-bonding in octahedral complexes by overlapping sidewise with π-orbitals of the same six ligands that are approaching the central metal ion from both sides along + x, – x, + y, – y + z and – z axes.

On the basis of the group theory, the metal orbitals are usually classified as

$dx^2 - y^2$, dz^2	e_g	or	Eg
s	a_{1g}	or	A_{1g}
p_x, p_y, and p_z	t_{1u}	or	T_{1u}
d_{xy}, d_{yz} and d_{zx}	t_{2g}	or	T_{2g}

In the above table, it is to be remembered that.

(i) a or A orbitals are always non–degenerate.

(ii) e or E orbitals, are double degenerate.

(iii) t or T orbitals are triple degenerate.

(iv) g expresses the gerade nature of orbitals and

(v) u expresses the ungerade nature of orbitals.

An orbital is called a gerade orbital if on traversing equal distances in a straight line in opposite directions from the centre of symmetry, the sign of wave function does not change.

4

Effective Atomic Number Rule

INTRODUCTION

In 1893 Werner put forward a theory, called Werner's theory of coordination compounds to explain the structure and properties of Co(III) and Pt(IV) ammines. These ammines are given below:

Co(III) ammines : $CoCl_3.6NH_3$ forange yellow), $CoCl_3.5NH_3.H_2O$ (pink), $CoCl_3.5NH$) (violet), trans-$CoCl_3.4NH_3$ (green), cis-$CoCl_3.4NH_3$ (violet), $CoCl_3.3NH_3$ (blue-green).

Pt (TV) ammines : $PtCl_4.6NH_3$, $PtCl_4.5NH_3$, $PtCl_4.4NH_3$, $PtCl_4.3NH_3$, $PtCl_4.2NH_3$.

Different postulates of Wemer's coordination theory are given below:

1. **Two types of valency shown by the central metal atom in coordination compounds**

   In complex compounds, the central metal atom exhibits two types of valency viz. primary valenq (also called principal valency) and *secondary valency* (also called *auxiliary valency*). The metal atom always tends to satisfy both of its valencies.

2. **Characteristics of secondary valency**

(i) In modem terminology, the secondary valency of the metallic atom in a complex compound is equal to the coordination number of that metal.

(ii) Secondary valency of a metal is satisfied either by anions or by neutral molecules alone or by anions and neutral molecules both. For example.

   (a) In $4KCN.Fe(CN)_2$ or $K_4[Fe(CN)_6]$, the secondary valency of Fe-atom is six and is satisfied by six CN^- ions (anions).

(b) In $CoCl_3.6NH_3$ or $[Co(NH_3)_6]Cl_3$, secondary valency of Co-atom is six and is satisfied by six NH_3 molecules (neutral molecules).

(c) In $CoCl_3.5NH_3$ or $[Co(NH_3)5Cl]Cl_2$, the secondary valency of Co-atom is six and is satisfied by five NH3 molecules (neutral molecules) and one CF ion (anion).

According to modem concept of complex compounds, the species satisfying the secondary valency are called ligands.

(iii) While writing the structure of a complex compound, the species satisfying the secondary valency and the metal are-written inside the coordination sphere. For example in $CoCl_3.4NH_3$, since four NH_3 molecules and two CF ions satisfy the secondary valency of Co-atom, its structure is written as $[CoCl_2.4NH_3]Cl$ or $[Co(NH_3)4d_2]Cl$.

(iv) The secondary valencies have directional nature, since the species satisfying the secondary valency (*i.e.*, ligands) are directed towards the fixed positions in space.

(v) The number of species satisfying the secondary valency gives a definite geometry to the complex compound. For example, when the number of monodentate species (ligands) which satisfy the secondary valency of the metal in a complex compound is 2, 3, 4 or 6, the geometry of the complex compound would be linear, trigonal planar, tetrahedral or square planar and octahedral respectively In each of the four Co (III) ammines viz., $CoCl_3.6NH_3$, $CoCl_3.5NH_3$, $CoCl_3.4NH_3$ and $CoCl_3.3NH_3$, since the secondary valency of Co-atom is 6, each ammine has octahedral geometry.

(vi) The species satisfying the secondary valency cannot be obtained in the free state, when the aqueous solution of the complex compound undergoes ionisation. For example, when the aqueous solution of $CoCl_3.4NH_3$ or $[Co(NH_3)_4Cl_2Cl$ undergoes ionisation, four NH_3 molecules and two Cl^- ions which satisfy the secondary valency of Co-atom are not obtained in the free state as shown below:

$$[Co(NH_3)_4Cl_2]Cl \rightarrow [Co(NH_3)_4Cl_2]^+ + Cl^-$$

The above discussion can be concluded by saying that the species satisfying the secondary valency are not ionisable (non-ionisable) and hence cannot be precipitated.

(vii) The attachment of the species satisfying the secondary valency with the metal atom is shown by a solid line (–).

3. Characteristics of primary valency : (i) In modem terminology, the primary valency of the metallic atom in a complex'compound is equal to the oxidation state (or oxidation number) of that metal, *e.g.*, the primary valency of Co-atom in all the four Co(III) ammines is equal to +3.

(ii) Primary valency of a metal in a complex compound is always satisfied by onions, *e.g.*, the primary valency of Co-atom in each of the four ammines is equal to −3 and is satisfied by three Cl^- ions.

(iii) The anions satisfying the primary valency are written outside the coordination sphere while the anions which satisfy both the valencies (dual character) are written inside the coordination sphere. Thus, the species satisfyinig primary valency may be present inside and/or outside the coordination sphere.

(iv) The species satisfying both the valencies (*i.e.*, the species placed inside the coordination sphere) are directed towards specific directions in space and hence they have directional characteristics. The species which satisfy primary valency and are placed outside the coordination sphere have no directional characteristics.

(v) The species satisfying the primary valency do not give any geometry to the complex compound.

(vi) The species satisfying the primary valency can be obtained either completely or partially in their free state. when the complex compound undergoes ionisation in aqueous solution.

Examples (a) When the aqueous solution of $CoCl_3.6NH_3$ or $[Co(NH_3)_6]Cl_3$ undergoes ionisation, all the three Cl^- ions which satisfy primary valency of Co-atom are obtained in the free state.

$$[Co(NH_3)_6]Cl_3 \rightarrow [Co(NH_3)_6]^{3+} + 3Cl^-$$

(b) Although the primary valency of Co-atom in $CoCl_3..5NH_3$ or $[Co(NH_3)_5Cl]Cl_2$ is satisfied by three Cl^- ions, only two Cl^- ions are obtained in the free state, when the aqpeous solution of this ammine undergoes ionisation.

$$[Co(NH_3)_5Cl]Cl_2 \rightarrow [Co(NH_3)_5Cl]^{2+} + 2Cl^-$$

The above discussion shows that all or some of the species satisfying the primary valencies are ionisable and hence can be precipitated by means of a suitable precipitating agent. For example, in the above ammine two Cl^- ions can be precipitated as AgCl (white ppt.) by adding aqueous solution of $AgNO_3$.

$$[Co(NH_3)_5Cl]Cl_2 \rightarrow [Co(NH_3)_5Cl]^{2+} + 2Cl^-$$

$$2Ag^+ + 2Cl^- \rightarrow 2AgCl \ (\textit{white ppt.})$$

In other words in understanding of co-ordination compounds and their properties began with the work of *Alfred Werner*. In 1899, Werner proposed an interpretation of co-ordination compounds which emphasized on the number and nature of groups attached to the metal ion.

This theory replaced the older concepts of *Berzelius* (*1819*), *Graham* (*1837*), *Claus* (*1856*). *Blomstrand (1869) and Jorgensen* (*1878–1894*) and became a fundamental part of the electronic theory of valency formulated by *G. N. Lewis* (*1916*).

According to Werner,

(a) Metals possess two types of valencies, so called :

   (i) Principal or ionisable valency, and

   (ii) Secondary or non-ionisable valency.

(b) Primary valencies are those which a metal exercises in the formation of its simple salts. Thus primary valencies of Pt, Co, Cu and Ag in the formation of their simple salts, *e.g.*, $PtCl_4$, $CoCl_3$ and $CuSO_4$ and AgCl are 4, 3, 2 and 1 respectively.

(c) Secondary valencies are the valencies which a metal cation exercises towards a neutral molecule or an anion in the formation of complex ions. The secondary valency is also termed as the coordination number of the metal cation under consideration.

   Thus, co-ordination number gives the number of neutral molecules or onions which may be linked to the metal cation in the formation of its complexes. For example, cobalt (III) and platinum (IV) were recognised as having six secondary valencies (co-ordination number = 6), whereas copper (I) has four (co-ordination number = 4).

(d) Primary valencies are satisfied by negative ions whereas secondary valencies may be satisfied by either negative groups or neutral molecules.

(e) Secondary valencies are directed in space and hence such compounds are capable of exhibiting the phenomenon of isomerism.

On the basis of this theory, Werner assigned to $CoCl_3.6NH_3$ the structural formula *by dashed lines.*

In this formula, size ammonia molecules are attached to the cobalt by secondary valencies where as three chlorine atoms are held to the cobalt by means of primary valencies. *Primary valencies are designated by solid lines and secondary valencies by dashed lines.*

Experimental Evidence in Support of Werner's Theory

A large number of complexes were studied by Werner and other workers, Werner used molar conductance values to determine the number of ions per molecule thus assigning groups to the first co-ordination sphere. Chemical methods like precipitation of Cl^- by $AgNO_3$ to determine the number of ionisable CF ions per molecule were also used. An interesting example is being discussed below :

Cobalt (III) ammine complexes : When an aqueous cobalt (II) chloride solution is treated with ammonia and then oxidised by air, the following six compounds are isolated :

(a) Luteo cobaltic chloride, $CoCl_3._6NH_3$: This is an orange yellow crystalline compounds. Its chief characteristics are :

(i) *Treatment of this compound in the solid state with sulphuric acid liberates all the chlorines as hydrogen chloride and leaves a sulphate. $Co_2(SO_4)_3.12NH_3$. Treatment of its solution with silver nitrate precipitates all the chlorines immediately. Thus, the bonding between cobalt and chlorine is purely ionic.*

(ii) *Treatment of solid with hydrochloric acid even at 100°C affects no removal of ammonia. Similarly, treatment of the solid with moist silver oxide gives a water soluble compound of composition $CoOOH.6NH_3$ which on addition of acids reforms the salts of type*

From the above experimental facts it follows that all the six ammonia molecules are unionisable and are held to the metal ion Co^{3+} in the first or inner coordination sphere whereas the ionisable Cl^- ions are in outer coordination sphere. From the above considerations, Werner formulated this complex as:

$$[Co(NH_3)_6]Cl_3$$

Conductance measurement indicates the presence of four ions, *i.e.*, one cobalt and three chloride ions (ammonia molecules are neutral). This is in accordance with the dissociation,

$$[Co(NH_3)_6 Cl_3 \rightleftharpoons \underset{\text{Cmplex ion}}{[Co(NH_3)_6]^{3+}} + 3Cl^-$$

(b) Roseo cobaltic Chloride $CoCl_3.5NH_3.H_2O$: This is a pink crystalline compound with the following characteristics :

(i) With $AgNO_3$, all the chlorines are precipitated as chloride from the solution of this complex indicating their ionizable character.

(ii) The water present in compound is firmly held at room temperature and is lost only at 100°C or above, showing its presence in the inner coordination sphere.

(iii) On treating with cone. H_2SO_4 it yields a compound $Co_2(SO_4)_3$ $10NH_3$. $5H_2O$ which on slight heating loses three water molecules, forming $Co_2(SO_4)_3.10NH_3.2H_2O$.

The above facts clearly indicate that water molecules and five ammonia molecules are firmly bound with cobalt whereas chlorines are loosely bound with cobalt. Thus, the formula of complex is

$$[Co(NH_3)_5.H_2O]Cl_3$$

Conductance measurements indicate the presence of four ions per molecule in very dilute solutions. This is in accord with the dissociation.

$$[Co(NH_3)_5.H_2O]Cl_3 \rightleftharpoons [Co(NH_3)5.H_2O]^{3+} + 3Cl^-$$

(c) Purpureo cobaltic chloride $CoCl_3.5NH_3$: It is a violet coloured compound with the following characteristics :

(i) When treated with cone. H_2SO_4, two chlorine atoms are lost as HCl, forming $Co(Cl)SO_4$.

(ii) Only two chlorines are precipitated immediately as AgCl on adding silver nitrate to its solution.

The above mentioned facts indicate the one chlorine atom is different from the rest two chlorine atoms. It means that only two of the three chlorines are ionisable as Cl^- ions.

On the basis of the above facts, Werner postulated the following formula : $[Co(NH_3)_5CL]Cl_2$

Conductance measurements indicate the presence of three ions per molecule in very dilute solutions. This is in accord with the dissociation.

$$[Co(NH_3)_5Cl]Cl_2 \rightleftharpoons [Co(NH_3)_5Cl]^{2+} + 2Cl$$

(d and e) Praseo-cobaltic chloride, violeo cobaltic chloride. $CoCl_3.4NH_3$: Two related compounds, one violet in colour (violeo-cobaltic chloride) and the other green (praseo-cobaltic chloride), have been prepared. Both have the same composition $CoCl_3.4NH_3$ but the chemical behaviour of these two compounds is quite different.

The conductances of solutions of these compounds show the presence of only two ions and from them only one chlorine is immediately precipitated by silver nitrate. Thus, from structural point of view their composition may be written as, $[Co(NH_3)_4Cl_2]Cl$

However, there were distinct differences in chemical behaviour between the two. Thus, although, only one third chlorine gets precipitated immediately on the addition of silver nitrate, second and third choleric atoms per molecule get also precipitated slowly and after long standing and no boiling. Werner explained this by assuming that the chlorines of inner co-ordination sphere are slowly replaced by water molecules :

$$[Co(NH_3)_4Cl_2]^+ + H_2O \longrightarrow [Co(NH_3)_4H_2O.Cl]^{2+} + Cl^-$$

$$[Co(NH_3)_4H_2O.Cl]^{2+} + H_2O \longrightarrow [Co(NH_3)_4.2H_2O]^{3+} + Cl^-$$

However, the rate of precipitation of these chlorines in these compounds is different.

Werner explained the above facts by assuming that praseo-cobaltic chloride is trans- type whereas violeo-cobaltic chloride is cis- type.

(f) $CoCl_3.3NH_3$: This is a blue green compound having the following characteristics :

(i) *It does not give any precipitate immediately on addition of silver nitrate to its solution (absence of ionisable chlorine).*

(ii) *Its dilute solutions did not conduct electricity at all, showing the absence of any ions in solution.*

From the above facts Werner came to the important conclusion that all the groups are linked to the central metal ion by unionisable secondary valencies and gave the following formula : $[Co(NH_3)_3Cl_3]$

Werner applied his theory to the above ammonia compounds of cobalt (III) chloride. The compounds may be formulated as :

In all the compounds discussed above, the primary valencies are designated by solid lines and secondary valencies by dotted lines. In each case, satisfaction of six secondary valencies is essential. Werner's theory is successful in

(i) Predicting the correct number of ions in each case.

(ii) Firm association of chlorine with cobalt in the last three cases and

(iii) The existence of two forms of $CoCl_3.4NH_3$ is explained. It is due to difference in three dimensional arrangement of four ammonias and two chlorines about the central cobalt (III) ion.

In order to differentiate, materials held by secondary valencies are enclosed in square brackets when writing formulae whereas materials held by primary valencies outside the brackets. Thus the compounds just considered are written as.

$[Co(NH_3)_6]Cl_3)$, $[Co(NH_3)_5(H_2O)]\,Cl_3$, $[Co(NH_3)_5Cl]Cl_2$
$[Co(NH_3)_4Cl_2]\,Cl$ and $[Co(NH_3)_3Cl_3]$

Criticism of Werner's Theory : One must appreciate the remarkable work done by Werner. He did this work 70 years ago and inspite of advances in theory and the increase in the number of coordination compounds, his postulates are as valid today as when they were presented over 70 years ago.

ELECTRONIC INTERPRETATION OF CO-ORDINATION

Sidgwick and Lowry gave an electronic interpretation of Werners theory.

According to Sidgwick, *Werner's primary valencies were regarded as amounting to electron transfer and the secondary or nonionic valencies to electron pair sharing.* An important point is to be noted that *all neutral molecules or onions which were capable of being coordinated to central metal ion have atoms with at least one unshared lone pair of electrons in their valency shells.* Thus, the linkage of these coordinating groups to metal ion by secondary valencies arises due to the donation of the electron pairs to the central metal by these groups. *Bonds of this type are termed as co-ordinate bonds. These bonds have been indicated by arrows.* For example the structure of $[Co(NH_3)_6]^{1+}$ ion may be written.

Effective Atomic Number (EAN): Stability of the complex compounds is explained by the fact that central metal atom acquires the effective atomic number of the next inert gas. *The effective atomic number is defined as the number of electrons of the central metal in a complex including those gained by bonding.*

In most of the cases the effective atomic number is the same as the atomic number of the next inert gas. However, there are a few cases, where the effective atomic number is more or less by one or two units as compared to next inert gas. EAN of some metals in certain complexes are given Table 1.

From the Table 1, it is evident that tendency to attain an inert gas configuration is a significant factor but not a necessary condition for complex formation. Some complexes of chromium and platinum are equally stable, yet the effective atomic number is short of inert gas configuration. For complex formation, it is essential to produce a symmetrical structure, *i.e.*, tetrahedral, square planar, octahedral, without any consideration of the number of electrons involved.

Table 1 : Effective atomic number of some metals in complexes

Atom	*Atomic number*	*Complex ion*	*Electrons lost in ion formation*	*Electrons gained by coordination*	*E.A.N number*	*Atomic of next inert gas*
Fe	26	$[Fe(CN)_6]^{4-}$	2	12	26–2+12=36	36 Kr
Co	27	$[Co(NH_3)_6]^{3+}$	3	12	27–3+12=36	36 Kr
Ni	28	$[Ni(CO)_4]$	0	8	28–0+8=36	36 Kr
Cu	29	$[Cu(CN)_4]^{3-}$-	1	8	29–1+8=36	36 Kr
Pd	46	$[Pd(NH_3)_6]^{4+}$	4	12	46–4+12=54	54 Kr
Cr	24	$[Cr(NH_3)_6]^{3+}$	3	12	24–3+12=33	36 Kr
Pd	46	$[PdCl_4]^{2-}$	2	8	46–2+8=54	54 Xe
Pt	78	$[Pt(NH_3)_4]^{2+}$	3	8	78–2+8=84	86 Rn

Objection. Sidgwick's electronic interpretation of co-ordination is not entirely satisfactory. Some objection are :

(i) The donation of electron pairs to a central cation would produce an improbable accumulation of negative charge on this ion. In the case of $[Co[NH_3)_6]^{3+}$ the donation of electron pairs from the ammonia molecules to the cobaltic ion would result an accumulation of negative charge on cobaltic ion. In this way cobaltic ion would become negatively charged with respect to the ammonia molecules.

(ii) Another difficulty presented by this concept is the fact that the electron pair, available for donation in water, ammonia and many other neutral molecules, is $2s^2$ pair. The electrons have no bonding characteristics and to excite them to a higher energy level would require more energy than is usually available in bond formation.

PAULING'S THEORY

Pauling introduced in 1931, a new theory which was based on the revolutionary idea of hybridisation. His theory was able to account for and even predict the geometrical shapes of many complexes. According to him.

(i) Strong bonds are not formed by pure s-or pure p- or pure d-orbitals.

(ii) Strong bonds are generally formed by hybridised orbitals formed by a suitable combination of orbitals. For example, the combination of s-orbital and p orbitals gives four equivalent orbitals of equal energy known as hybrids. This type of combination is known as sp^3 hybridisation. In complexes, other types of hybridisation like d^2sp^3, sp^3d^2 and dsp^2 are more common.

(iii) In transition metal atom, d-orbitals are, partially filled up and hence it is the number of d-electrons in the central atom (forming complex) that determines the possible modes of hybridization and the shapes of complex.

(iv) In complex formation, Hund's rule of maximum multiplicity is strictly followed. However, in complex formation redistribution of electrons and complete pairing of electrons may take place. In this way some orbitals are vacated and made available for hybridisation.

Table 2

Bond Hybridisation	*Geometrical Shape*	*Typical Example*
d^2sp^3	Octahedral	$Fe(CN)_6^{3-}$
sp^3d^2	Octahedral	FeF_6^{3-}
sp^3	Tetrahedral	$Zn(NH_3)_4^{2+}$
dsp^2	Square planar	$Ni(CN)_4^{2-}$

Pauling's theory provides a suggestive explanation of the experimental facts and, in particular, of the magnetic susceptibilities of co-ordination compounds. Pauling's theory also differentiated between complex ions in which the bonding could be considered as ionic and those in which it was purely covalent.

Let us apply this theory to various examples;

Octahedral complexes : Octahedral complexes result from either sp^3d^2 or $d^2\,sp^3$ hybridisation as in the following examples.

(a) Ferricyanide ion, $[Fe(CNH)_6]3^-$: The structures of the Fe atom and the Fe^{3+} ion are given at (i) and (ii).

The five unpaired electrons in the Fe^{3+} give a calculated magnetic moment of 5.91 magneton. The magnetic moment for the $Fe(CN)_6^{3-}$ ion, however, is 2.3 suggesting, perhaps, one unpaired electron.

In order to explain the magnetic moment of $Fe(CN)_6^{3-}$, Pauling suggested that the unpaired *3d* electrons in the Fe^{3+} undergo pairing to give the arrangement as shown in (ii) of Fig. 1.8. The resulting complex would have one unpaired electron, and the d^2sp^3 orbitals are available for hybridisation.

The electrons, to occupy the d^2sp^3 hybrid orbitals, would be provided by the lone pairs from each of the six cyanide ion ligands. This means that the bonds between the ligands and the central ion are being coordinated linkages. Complexes like $|Fe(CN)_6]^{4-}$ are known as *inner-orbital complexes* because the *d-orbitals* involved are from a lower shell than the *s* and p-orbitals.

(b) Ferrocyanide ion $[Fe(CN)_6]^{4-}$: In the ferrocyanide ion there is one more electron than in the ferricyanide ion. This extra electron pairs with the unpaired 3d-electron in (iii), to give the structure shown in

Fig. 1.8. The ferrocyanide ion has, therefore, no unpaired electrons and would be expected to be diamagnetic, this is found to be so. It is octahedral with d^2sp^3 hybridisation.

Other examples resembling structures like the ferri , or cyanide ions include $Mn(CN)_6^{3-}$, $Cr(NH_3)_6^{3+}$ and $Cr(CN)_6^{4-}$.

(c) Ferrifluoride ion $[FeF_6]^{3-}$: The $[FeF_6]^{3-}$ is octahedral like the ferri- and ferrocyanide ions. However, it has a magnetic moment of 6.0 which is in close agreement with the value of 5.9 – a value which would be expected from a structure with five unpaired electrons.

In order to explain this, it is assumed that no pairing of electrons in Fe^{3+} ions takes place and sp^3d^2 orbitals are used for hybridisation.

The FeF_6^{3-} is an example of an outer orbital complex because the d-orbitals involved in the hybridisation are from a higher shell than the s- and p-orbitals.

Other similar outer-orbital complexes are CoF_6^{3-} , $Fe(NH_3)_6^{2+}$: $Ni(NH_3)_6^{2+}$ etc.

Tetrahedral Complexes : Tetrahedral complexes result from sp^3 hybridisation. Complexes of Zn^{2+} are tetrahedral because they involve sp^3 hybrid bonds. The electron arrangements in the Zn atom and Zn^{2+} ion are shown at (i) and (ii) below :

In the Zn^{2+} ion, all the $3d$ orbitals are fully occupied and therefore, they cannot take part in bond formation. However, sp^3 hybridisation takes place and four sp^3 hybrid orbitals are formed. The electrons to occupy sp^3 hybrid orbitals would be provided by the lone pairs from each of the four ammonia molecules in $Zn(NH_3)_4^{2+}$. Thus, $Zn(NH_3)_4^{2+}$ will be tetrahedral. As there are no unpaired electrons, $Zn(NH_3)_4^{2+}$ will also be diamagnetic.

Other similar tetrahedral complexes are formed by Cd^{2+}, Hg^{2+}, Cu^+ and Ag^+ ions; all of which have the same structure as Zn^{2+}.

Square planar complexes : Square planar complexes result from dsp^2 hybridisation. This type of hybridisation involves mixing of one *d*-, one *s*- and two $2p$ orbitals, resulting in the formation of four equivalent hybrid orbitals which are planar and mutually at right angles pointing towards the four corners of a square. The d–orbital involved is the dx^2–

y^2 orbital in the shell lower than that of the *s*- and p-orbitals. Square planar complexes tend to be formed when the central ions have only one d-orbital available in the lower shell.

(a) Nickelocyanide ion $Ni(CN)_4^{2-}$: The electronic arrangements for the Ni atom and the Ni^{2+} ion are shown below :

If the two unpaired 3*d* electrons couple, the structure will be as given in (iii) so that dsp^{2-} hybrid bonds can be formed. All the electrons are paired so that the $Ni(CN)_4^{2-}$ ion is diamagnetic. Other similar examples are $PtCl_4^{2-}$, $PdCl_4^{2-}$, etc.

(b) Cuprammonium ion : $Cu(NH_3)_4^{2+}$: It has a coplanar square structure and is paramagnetic with a magnetic moment of 1.8. A square planar structure is obtained by *dsp*^{2} hybridisation, keeping one electron unpaired. Such facts are accounted for on the basis of the structures shown at (i), (ii) and (iii).

Objections to the Shape of Cuprammonium Ion

(a) The necessity to promote one electron from a 3*d* into a *4p* level is not satisfactory. The 4p electrons would be expected to be easily lost which would mean that the complex could be easily oxidised, but this is not so.

(b) Placement of the odd election in the 3*d* level as in (iv) would suggest *sp*^{3} hybridisation with a resulting nonexistent tetrahedral structure.

Objections to Pauling's Theory : Pauling's theory explained the chemistry of co-ordination compounds satisfactorily but failed to explain some of the aspects :

(a) The peculiar behaviour of ions of divalent nickel, palladium, platinum and trivalent gold which apparently do not form the expected 5–co-ordination complexes.

(b) Pauling's theory proposed a form of hybridisation (*dsp*^{2}) suitable for square-planar structures but this does not explain why this structure is preferred to other possible stereochemical configurations, such as tetrahedral or 5–co-ordinated trigonal bipyramid which does, in fact, occur in iron carbonyl.

(c) Pauling's ideas offered a little or no explanation to the absorption spectra of complexes or of their relative stabilities or reactivities.

(d) It is well known that both the colour and magnetic moment of transition metal complexes are due to their possessing d-orbital electrons.

It means that there should be a quantitative connection between spectra and magnetic moment. Unfortunately this connection is not revealed by Pauling's theory.

(e) It cannot explain why certain complexes are more labile than the others.

Labile complexes are those in which one ligand can be easily displaced by another ligand. On the other hand, inert complexes are those in which displacement of ligands is slow.

(f) It does not take into consideration the splitting of d–energy levels.

(g) It does not account for the relative energies of different structures.

ELECTRONEUTRALITY PRINCIPLE

Earlier while discussing the Sidgwick's interpretation of Werner's compounds we have shown that coordinate bonds are formed by the donation of a pair of electrons by each ligand atom to metal ion. Once the bond is formed there is no ditinction between a coordinate bond or a covalent bond for in both the cases the electrons are equally shared.

Thus, in $[Co(NH_3)_6]^{3+}$ complex, out of 12 electrons shared between cobalt (III) and 6N atoms of ammonia molecule, the metal gains 6 electrons and so the formal charge changes from +3 to –3. Thus, the formation of coordinate bonds implies the accumulation of negative charge on the metal atom.

But Pauling has shown that such a situation need not arise which he expressed in his *electroneutrality principle.* According to this principle *the electrons are distributed in a molecule in such a way as to make the residual charge on each of atom zero or very likely zero except that hydrogen and the most positive metals can acquire partial positive charge and the most electronegative atoms can acquire partial negative charge.*

Thus obviously it implies that the M–L bond is not perfectly covalent. In other words, the bond is partially ionic and that the electron pair is attached more strongly to the ligand atom. Thus, if it is supposed that the M–L bond is 50 percent ionic, then the metal would be exactly neutral and the charge on each ligand atom would be +1/2. The formation and stability of aquo and ammonia complexes can be explained on the basis of ion dipole electrostatic attraction but for ligands having negative character electroneutrality principle has to be invoked to explain their stability.

In addition Pauling has also suggested that accumulation of negative charge can be avoided if it is assumed that in addition to o bond, the central metal atom also forms n bonds by the back donation of electrons to the ligand atoms. Thus, the formation of

$$M \xrightarrow[\sigma]{\pi} L$$

π bond results in the decrease of negative charges on the metal atom. The *n* bond formation may take place in two ways :

(i) ***dπ–pπ bonding*** : *The first atom in a group which acts as donor atoms, (e.g., N,C in NO_2^- and CN^- respectively) do not have d-orbitals available for bonding. So the electrons donated by the filled dn orbitals (dxy, or dyl orbitals) are accommodated in the vacant p orbitals of the ligand atom. Thus, there is an effective dπ–pπ overlapping between metal and ligand orbitals.*

(ii) ***dπ-dπ bonding*** : *In this type of bonding the back donated electrons from dπ orbitals of the metal are accommodated in the dπ orbitals of the donor atoms. Such donor atoms, naturally will have vacant dn orbitals and belong to the second short period or of the later periods e.g., P, S, As in PR_3, SR_2 and AsR_3, respectively.*

The formation of π bonding not only avoids the accumulation of negative charges on the metal atom but also strengthens the σ bond which explains the stability of many transition metals complexes with very poor electron pair donor atoms, such as CO, PX_2, PR_2, SR_2, and

CRYSTAL FIELD THEORY

Introduction : This theory which was first introduced by Bethe and Van Vleck and applied mainly to ionic crystals, and is, therefore, called **Crystal Field Theory** (1956). Later **Orgel** was largely responsible for developing its general chemical aspects. The crystal field theory is mainly concerned with the interaction of d-orbitals of central metal with the surrounding ligands that produce crystal field effects. The crystal field theory is based on the following assumptions:

(i) The metal ion and ligands act as point charges and the interaction between them is purely electrostatic, *i.e.*, metal-ligand bonds are 100 per cent ionic.

(ii) There is no interaction between metal orbitals and ligand orbitals.

(iii) The d–orbitals on the metal which were all of the same energy, *i.e.* degenerate in a free atom, have their degeneracy destroyed by the ligands when a complex is formed.

Salient Features of Crystal Field Theory

The various postulate are summarised below

(a) A complex is considered to be a combination of central metal ion surrounded by various ligands.

(b) The ligands in the complexes to be considered are either negatively charged ions *e.g.*, F and CN^- or neutral molecules, *e.g.* H_2O and NH_2, which have a dipole such that the negative ion is closest to the central ion.

(c) The attraction between central metal and ligand in a complex is regarded as purely electrostatic. The interaction between the electrons of the cation and those of the ligands is purely repulsive. It is these repulsive forces that are responsible for causing the splitting of the d-orbitals of the metal cation into two groups, t_{2g} and *eg*. The splitting depends on whether the ligands are arranged in an octahedral, tetrahedral or squareplanar way around the central ion. This effect is known as *crystal field splitting.*

(d) The electrons of the metal ion are repelled by the negative field of the ligands and therefore the metal electrons will occupy those d-orbitals which have their lobes farthest way from the direction of the ligand.

(e) In this theory, the covalent character of bond between metal and ligand is not taken into account.

(f) This theory does not permit the electrons to enter the metal orbitals. It means that it does not consider any orbital overlap.

(g) The number of ligands and their arrangement around the central ion will determine the crystal field.

Crystal field splitting of d-orbitals : In an isolated gaseous metal ion, all the five d orbital are degenerate, *i.e.*, have the same energy. The electrons which occupy such orbitals, do so according to the Hund's rule of maximum multiplicity.

Crystal field theory is basically concerned with the effect of different arrangement of surrounding ligands on the energy of the d-orbitals, and the subdivision of the orbitals into two groups, t_2g and e_g.

On the approach of the ligands, the electrons in the d-orbital of the central ion are repelled by the lone pairs of the ligands. This repulsion will raise the energy level of the d-orbitals.

All the ligands approaching the energy of each orbital will increase by the same amount. In other words they will remain degenerate, although they will have now higher energy than before.

But actually the d-orbitals differ in their orientation. Thus the energy of the orbitals lying in the direction of the ligands is raised to a larger extent than that of the orbitals lying in between the ligands. It means that the five degenerate d-orbitals of the metal ion will split up into two sets of orbitals having different energies. This splitting of five degenerate d-orbitals of the metal ion under the influence of approaching ligands, into two sets of orbitals having different energies, is called *crystal field splitting or energy level splitting.*

The ligand field splitting depends on whether the ligands are arranged in an *octahedral, tetrahedral or square planar* way around the central ion.

The magnitude of the crystal field splitting of the d-orbitals energies may be represented by the symbol Δ. It is used to denote energy difference (the total energy difference may be regarded as made up of the energy lost or released in the lower levels together with the energy gained in the upper levels). The crystal field splitting energy, Δ, is frequently measured in terms of a parameter, Dq. The magnitude of splitting is arbitrarily set at 10 Dq, *i.e.*, 2 Dq for each unpaired d–electron.

Crystal Field Splitting by Different Geometrically Arrangements of Ligands

The way in which the degeneracy of d-orbitals is resolved is different in octahedral, tetrahedral and square planar fields. We shall discuss separately the splitting in different fields.

Octahedral complexes : In the octahedral complex, the six ligands are arranged octahedral around a central metal.

In such an arrangement as the $dx^2–y^2$ and dz^2 orbitals lie along the *x, y* and *z* axes and so point directly towards the ligands, they experience much more repulsion than the remaining d-orbitals (*dxy*, *dyz* and *dzx*) which are directed in between the or, y and z axes.

Consequently, the energy of $dx^2 - y^2$ and dz^2 orbitals is increased much more in comparison to other d-orbitals. Thus, the d-subshell is splitted up into two degenerate sets, one consisting of more stable (lower energy) orbitals *dxy, dyz* and *dzx* and the other, less stable higher energy) orbitals $dx^2 - y^2$ and dz^2. These are commonly referred to as t_{2g} and e_g or d_ϵ and d_γ sets of orbitals respectively.

Degeneracy of all the five d–orbitals in the isolated central ion is represented by state I. The state II represents degeneracy (hypothetical) of all the five orbitals at a higher energy level. The state III represents *crystal field splitting.* The crystal field splitting is thus the energy difference between t_{2g} and e_g orbitals. It is denoted by the symbol Δ_0 or 10 Dq. (The subscript 0 indicated octahedral complexes).

The geometry of an octahedral system shows that the energy of the t_{2g} orbitals is 2.5 Δ_0 or 0.4 Δ_0 or 4 Dq less than that of the hypothetical degenerated d orbitals which results in crystal field splitting is ignored. The energy of the e_g orbitals is 3/5 Δ_0 or 0.6 Δ_0 or 6 Dq above that of the hypothetical degenerate d–orbitals.

An electron prefers to move into an orbital of a lower energy. It is obvious that if an octahedral complex contains one d-electron, this electron will occupy one of the t_2g orbitals.

The t_2g orbital has an energy 4 Dq less than that of the degenerate orbitals. Thus the complex will be 4 Dq more stable than predicted by the purely electrostatic theory which does not recognise any splitting of energy levels. This value, 4 Dq, is called the *crystal field stabilising energy* (C.F.E) of the complex under discussion.

If for each electron entering into a tag orbital, the crystal field stabilising energy is 4 Dq. then for each electron entering into an e_g orbital, the *destabilising energy is* 6 Dq. In other words, the stabilization energy assigned for an electron in e_g orbital is – 6 Dq. Working on this basis, it is possible to calculate crystal field stabilization energies for various metal ions in octahedral complexes.

Distribution of electrons in *d* orbitals and CFSE of octahedral complexes : The distribution of electrons in the *d* orbitals is based upon the following rules :

(i) *Electrons prefer to occupy the orbitals of lowest energy.*

(ii) *Electrons obey Hund's rule, i.e., electrons pairing will not take place until all the available orbitals of a given set contain one electron each.*

(iii) *If the energy to place an electron in the upper* e_g *level is greater than the energy to pair electrons (pairing energy = P) in the lower* t_2g *level* (10 Dq < P), *then the electrons will pair up in* t_2g *orbitals.*

(iv) *If the energy to place an electron in the upper eg level is less than the pairing energy in the lower* t_2g *orbital* (10 Dq < P), *then the electrons will occupy* e_g *orbitals.*

Weak and strong field cases : Ligands which cause only a small degree of splitting of *d* orbitals, *i.e.,* weak field, are called *weak field ligands* and those which cause a large splitting. *i.e.,* strong field, are called *strong field ligands.* The magnitude of 10 *Dq* and hence the distribution of the electron in *d* orbitals are very much influenced by the strength of the ligand field. The CF splitting ability of the ligands decreases in the following order :

$$CO > CN^- > NO_2^- > en > NH_3 > H_2O > F^- > OH^- > Cl^- > Br^- > I^-$$

Strong field ligands ... Weak field ligands

In a weak field, 10 Dq is generally less than pairing energy, *i.e.,* 10 Dq < P. It means that electrons obey Hund's rule and remain unpaired. But in the case of a strong field, 10 Dq > P; it follows that electrons pair up in the lower energy orbitals. We shall now consider the distribution of *d* electrons in octahedral complexes, in the presence of weak field and strong field ligands.

In octahedral complexes with d^1 configuration, the *d* electron will occupy the lower t_{2g} orbitals irrespective of the field strength and the crystal field stabilisation energy is – 4Dq. For d^2 configuration, the two electrons will occupy the lower two t_{2g} orbitals and CFSE is – 4Dq × 2 = 8 Dq. Similarly for d^3 configuration CFSE is – 12Dq and the t_{2g} level is just half-filled. Thus complexes with d^1, d^2 and d^3 configuration, irrespective of the ligand field strength, remain paramagnetic, with one, two and three unpaired electrons respectively.

For d^4 configuration, two possibilities will arise, that is, the fourth electron can either enter the upper e_g level or pair up in the lower t_{2g} level. Which of these two arrangements occurs depends upon the values of 10 Dq and P, which in turn depends upon the strength of the ligands.

In a *weak field,* 10 Dq < P, hence the fourth electron prefers to enter one of the e_g orbitals rather than pairing in t_{2g} orbitals. This way, it loses somewhat less energy in the form of CFSE by entering the destablised e_g level. The net CFSE energy is :

$$\text{CFSE} = (3 \times -4\text{ Dq}) + 1 \times (+6\text{Dq}) = -6\text{Dq}$$

In a *strong field* where 10 Dq > P, the fourth electron prefers to pair up in one of the three lower t_{2g} orbitals rather than entering the upper e_g orbitals. CFSE for this type of complex is (4 × – 4Dq + P) because all the four electrons enter t2g orbitals and some energy is used to pair up the electron.

Thus, for d^4 configuration the electronic arrangement' differs in weak and strong fields. In weak field there are four unpaired electrons while in the strong ligand field, there are only two unpaired electrons. These two arrangements thus differ in the number of unpaired electrons. The arrangement with maximum unpaired electrons is called *'high spin'* or *'spin free'* and the one with minimum number of unpaired electrons is called *low spin or spin paired* arrangement.

For d^5 configuration, in a weak field three electrons occupy the t_{2g} orbitals and the other two in two e_g orbitals (one in each). CFSE is zero because the presence of two electrons in the unfavourable e_g level exactly balances, the stabilisation from the three electrons in the t_2g level.

In other words, the half filled shell (d^5) is physically symmetrically and no stabilisation can occur through the application of external ligand field. In case of *strong field* the fifth electron will pair up in the t_{2g} orbital.

$$\therefore \qquad \text{CFSE} = -20\text{ Dq} + 2\text{P}$$

For d^6 configuration in a weak field, the sixth electron will pair up in the lower t_{2g} orbitals because all the five d-orbitals are already half-filled CFSE = – 4Dq + P. In case of *strong field,* the sixth electron will pair up in the t_{2g} orbitals. Thus, all the three t_2g orbitals will have paired electrons.

$\therefore$ CFSE = – 24 Dq + 3P

Similarly, *for* d^7, d^8 and d^9 and d^{10} configurations, the d–electrons of the metal can be distributed and their CFSE can be calculated.

It is clear that for d^1, d^2, d^3, and d^8, d^9, d^{10} configurations, the distribution of electrons and CFSE are the same irrespective of the strength of the ligand field. But for configurations d^4 to d^7, two arrangements (high-spin and low- spin) are possible depending upon the strength of the ligand field.

Tetrahedral complexes : In tetrahedral complexes, the four ligands occupy the alternate corners of a cube, in the centre of which is placed the metal cation. The lobes of the dxy, dyz and dzx orbitals point towards the middle of the edge of the cube while those of the $dx^2 - y^2$ and dz^2 lie on the bisectors of the angle between pairs of ligands and so pass through the centre of cube faces.

Obviously, none of the d-orbitals point directly towards the ligands. However, the calculations reveal that the *dxy dyz* and *dxz* (termed t_{2g}) are closer to ligands than $dx^2 - y^2$ and dz^2 (termed e_g). In the formation of a regular tetarahedral structure in a complex, the approach of ligands raises the energy of both sets of orbitals, but since the t_{2g} orbitals correspond more closely to the position of the ligands, their energy increases most as compared to e_g.

Thus, e_g orbitals (lower energy) are filled first. This is opposite to what happens in an octahedral complexes.

For maintaining the average constant, t_{2g} orbitals are raised to 4 Dq while e_g orbitals are lowered by 6 Dq from the barycenter. Thus, *the energy level scheme for the tetrahedral symmetry is exactly the reverse of that for octahedral symmetry.*

It has been observed that the energy separation between the e_g and t_2g levels Δ_1, in tetrahedral symmetry is smaller than in octahedral symmetry Δ_0 (10 Dq). This observation is as expected because the d-orbitals are not directly affected by the ligands and moreover the number of ligands is smaller. For equivalent ligand charges and distances:

$$\Delta_t = -\frac{4}{9} D^{\circ}$$

where the minus sign means the reversal of the orbital levels. Because of the low value of Dq in tetrahedral symmetry *crystal field favours the formation of octahedral complexes than tetrahedral complexes.*

Low spin (strong field) tetrahedral complexes have not yet been obtained, probably because of crystal field stabilisation energies are never sufficient to bring out spin pairing.

Thus. *only high spin* or *weak field* cases are to be considered for tetrahedral complexes. This simplifies the treatment of electron configuration and CFSE.

In a weak field, pairing energy (P) is more than Δ_t. Hence, electrons enter even the higher levels and t_{2g} remain unpaired until the sixth electron forces pairing. For example in a complex with d^4 configuration two electrons occupy the lower e_g orbitals and two occupy the higher t_{2g} orbitals, all the four electrons remaining unpaired.

Hence, CFSE = – 4Dq *i.e.*,

$$(2 \times -6Dq) + (2 \times 4Dq) = -4Dq$$

The distribution of electrons and the crystal field stabilisation energies for d^1 to d^{10} configuration are given in Table 1.6.

Table 3 : Distribution of D electrons for weak field in tetrahedral complexes.

d^n	Distribution of electrons e_g	t_{2g}	No. of unpaired electrons	CFSE
d^1	↑		1	– 6Dq
d^2	↑ ↑		2	– 12Dq
d^3	↑ ↑	↑	3	– 8Dq
d^4	↑ ↑	↑ ↑	4	– 4Dq
d^5	↑ ↑	↑ ↑ ↑	5	– 0Dq
d^6	↑ ↓ ↑	↑ ↑ ↑	4	– 6Dq + P
d^7	↑ ↓ ↑ ↓	↑ ↓ ↑ ↑	3	– 12Dq+2P
d^8	↑ ↓ ↑ ↓	↑ ↓ ↑ ↑	2	– 8Dq + 3P
d^9	↑ ↓ ↑ ↓	↑ ↓ ↑ ↓ ↑	1	– 4Dq + 4P
d^{10}	↑ ↓ ↑ ↓	↑ ↓ ↑ ↑ ↑ ↓	0	– 0Dq + 5P

(c) Square planar complexes : In the square planar complexes due to the absence of the ligand groups along the z-axis, the dz^2 orbitals becomes most stable and $dx^2 - y^2$ least stable.

However, most of the square planar complexes are such, where the two ligands along the z-axis are not totally removed but placed at a longer distance than the ligands along the x and y axes (*e.g.*, Cu^{2+} and Ni^{2+} complexes). In such cases the *dz* orbital though becomes more stable than $dx^2 - y^2$ and d_{xy} orbitals, it remains less stable in comparison to *dxz* and *dyz* orbitals, which still form a degenerate set. The sequence of orbitals in the increasing order of energy would then be *dxz* and *dyz*, dz^2, *dxy* and $dx^2 - y^2$.

The difference in the energy levels, created within the *d* subshell by the crystal field, are small and the electrons from lower levels can be excited to the higher levels, usually by the absorption of visible light. This phenomenon of easy *"electron Transition"* is responsible for the complexes appearing coloured.

Factors Affecting D

The magnitude of Δ or 10Dq is influenced by a number of factors such as :

(a) Geometry of the complex : In octahedral complexes, the splitting of *d* orbital has been found to be more than twice than in tetrahedral complexes, for the same metal ion and ligands. This difference in Δ value is due to the following facts.

(i) *In octahedral complexes six ligands are involved whereas in tetrahedral complexes only four ligands are involved. This results in* 33 *per cent decrease in the field strength, provided the other factors do not change.*

(ii) *In octahedral complexes ligands are positioned exactly in the direction of* d_z^2 *and* $dx^2 - y^2$ *orbitals whereas in tetrahedral complexes, the ligands are not positioned at any of the d-orbitals and thus have more influence on the* t_{2g} *orbitals than on the* e_g *orbitals. However, in case of square planar complexes, the degree of splitting is more than in a tetrahedral field.*

(b) Nature of the ligands : If we compare the absorption spectra of various transition metal complexes, we observe that the position of absorption spectra and hence the value of Δ or 10 Dq for any given metal ion depends upon the nature of the ligands attached. In order to illustrate this point, we have given two values of Δ_0 or 10Dq for a number of Cr^{3+} complexes with different ligands in Table 4.

Table 4

Complex	*Absorption peak cm^{-1}*	*10 Dq or Δ_0 (Kcal/mol)*
$[CrCl_6]^{3-}$	13640	38.96
$[Cr(H_2O)6]^{3+}$	17830	50.94
$[Cr(NH_3)_6]^{3+}$	21680	61.94
$[Cr(CN)_6]^{3-}$	262.0	74.08

From the above table we can draw the important conclusion that the cyanide ligand produces more splitting and is therefore a strong ligand whereas chloride ligand produces less splitting and is therefore a weak ligand. Similarly, we can compare the various ligands, and then arrange them in the order of their strength, in the form of a series called spectrochemical series.

$$I^- < Br^- < S^{2-} < SCN^- < Cl^- < NO^{3-} < F < OH^-$$
$$< H_2O < NCS^-, NH_3 < en < dipy < NO_2^- < CN^- < CO.$$

(c) Position of the metal in transition series : This in an important factor and the magnitude of crystal field splitting depends upon the position of metal in transition series, *i.e.,* whether it belongs to first, second or third series. In general, the value of Δ_0 or 10 Dq increases on descending a group of transition element, *i.e.,* from first to third transition series. This point has been illustrated in Table 5.

Table 5

Complex	*Δ_0 cm^{-1}*	*Δ_0 (Kcal/mol)*
$[Co\ (NH_3)_6]^{3+}$	23,000	65.70
$[Rh(NH_3)_6]^{3+}$	34,000	97.14
$[Ir(NH_3)_6]^{3+}$	41,000	117.1

(d) Charge on the metal ion : As earlier stated that crystal field theory is based on electrostatic model. It means that the ionic charge present on the central metal has a direct effect on the magnitude of 10Dq.

A generalisation is that a metal with higher charge produces more splitting than an ion with lower charge. For example, Δ_0 or 10 Dq for hexa-aquo complexes of Cr^{2+} and Cr^{3+} are 39.7 Kcal mol^{-1} and 51.94 Real mol^{-1}.

Measurement of Δ or 10 Dq

The splitting of d-orbitals by a ligand field into two sets with different energies provides a possible electronic energy change within a complex. An electron in an orbital of lower energy might be excited into an orbital of higher energy by the absorption of radiation. It so happens that the radiation necessary for such excitation lies in the visible region of light and this explains why many complexes are coloured.

Study of absorption spectra of complexes enables Δ values to be obtained. A band or bands, in the absorption spectra, can be related to the excitation of an electron or electrons, between d-orbitals splitted by a ligand field.

The ability to interpret such spectroscopic evidence is one of the advantages of crystal field theory. The situation in the $[Ti\ (H_2O)_6]^{3+}$ ion, hexaquotitanium (III) ion, is particularly simple for there is only one *d* electron (d^1) involved. In a ground state, it will occupy one of the t_{2g} (d_ϵ) orbitals but absorption of radiation will excite it into one the e_4 $(d\lambda)$ orbitals.

Table 6 : Crystal field spliting (kcal/mole).

Metal ion		*Ligand*	
Ti (III)	$3d^1$	H_2O	NH_3
V (III)	$3d^2$	58	—
Cr (III)	$3d^3$	51	—
Fe (III)	$3d^5$	50	—
Mn (II)	$3d^5$	39	—
Co (III)	$3d^6$	22	65
Fe (II)	$3d^6$	53	—
Co (II)	$3d^7$	30	29
Ni (II)	$3d^8$	28	31
Cr (II)	$3d^4$	24	43
		36	

The absorption spectrum of $[Ti\ (H_2O_6]^{3+}$ shows a strong band corresponding to wavelength of 5000A°. This represents an energy change of 57.15 Kcal/mole which must be the Δ value for this complex. Light of wavelength 5000A° is green. If such a light is absorbed, the transmitted light appears blue and red, *i.e.*, purple, and $[Ti(H_2O)_6]^{+3}$ ion is in fact, purple.

Table 6 gives a few of the values of Δ for the various transition metal ions and ligands. While value of Δ is approximately constant for ions of a given charge with the same ligand, changing the ligand does change A and thus alters the absorption spectrum associated with the metal ion. It is this change in the crystal field splitting of the d–orbital energies that is responsible for the colour change that occurs when one ligand is replaced with an other.

APPLICATIONS OF CO-ORDINATION COMPOUNDS

Profound changes occur in the properties of a metal ion on complex formation. These changes are generally reflected in solubility, colour, stability, magnetic properties, etc. These changes in properties find manifold applications.

1. Dyes : The characteristic colours of co-ordination compounds would indicate their uses as dyes. Alfred Werner demonstrated the relation of co-ordination with dyeing and he showed that several compounds which were capable of forming metal-chelate compounds were able to dye cloth pretreated with ferric hydroxide.

The first complete study of co-ordination compounds as dyes was given by G.T. Morgan and his co-workers in the early 1920.

2. As catalysts : Many enzymes, which serve as the catalysts in living system, are coordination compounds. For example, the decomposition of hydrogen peroxide is catalysed by many things, including iron compounds.

$$2H_2O_2 \xrightarrow{\text{catalyst}} 2H_2O + O_2.$$

This is evident from the following points

(i) *Ordinary hydrated ferric ion has a relative activity of unity,*

(ii) *A co-ordination compound of iron, the he me (human blood), has a relative activity of one thousand, and*

(iii) *Catalase, a heme surrounded by a complicated protein structure, has a relative catalytic activity of ten billion.*

3. Analytical Applications : The basis of qualitative detection of metal ions is the complexation reactions of such ions. Common examples of qualitative detection of metal ions by way of complexation reactions include detection of nickel (II), as red coloured bis (dimethylglyoximato) nickel (II), of cobalt as blue coloured tetrakis (isothiocyanato) cobalt (II), $[Co(NCS)_4]^{2-}$, of iron as blood red hexakis (isothiocyan-to) iron (III), $[Fe(NCS)]_3^-$ etc. These are as follows.

(a) Aluminium and zinc salt solutions form with excess of NaOH, soluble sodium aluminate and sodium zincate.

$$AlCl_3 + 3NaOH \longrightarrow \underset{ppt}{Al(OH)_3} + 3NaCl$$

$$Al(OH)_3 + NaOH \longrightarrow \underset{\text{Sod. alu min ate}}{NaAlO_2} + 2H_2O$$

$$ZnCl_2 + 2NaOH \longrightarrow \underset{ppt.}{ZnOH_2} + 2NaCl$$

$$Zn(OH)_2 + 2NaOH \longrightarrow \underset{\text{Sod. zincate}}{Na_2\,ZnO_2} + 2H_2O$$

Similarly, chromium hydroxide dissolves in excess of NaOH forming sodium chromite.

$$CrCl_3 + 3NaOH \longrightarrow Cr(OH)_3 + 3NaCl$$

$$2Cr(OH)_3 + 2NaOH \longrightarrow \underset{\text{Sod. chromite}}{2Na_2Cr_2O_4} + 4H_2O$$

Chromium salts in the presence of oxidising agents such as Br_2 water, H_2O_2, form a soluble yellow chromate ion.

$$2CrCl_3 + 10NaOH + O \longrightarrow \underset{\text{Sod. chromate}}{2Na_2CrO_4} + 6NaCl + 5H2O$$

These reactions have been utilised for the separation of aluminium from iron and chromium and zinc from manganese.

(b) Iodine is having feeble solubility in water. However, its solution is prepared by dissolving iodine in a solution of potassium iodide. This solubility is due to the formation of complex KI_3.

$$KI + I_2 \longrightarrow KI_3 \rightleftharpoons K^+ + I_3^-$$

When potassium iodide solution is added to mercuric chloride solution, a scarlet red ppt. of mercuric iodide is obtained which dissolves in excess of KI forming K_2HgI_4.

$$HgCl_2 + 2KI \longrightarrow HgI_2 + 2KCl$$

$$HgI_2 + 2KI \longrightarrow \underset{\text{(Complex)}}{K_2HgI_4}$$

When this complex is added to sodium hydroxide, the solution is known as Nessler's reagent which is employed for the *detection of ammonia.* It forms brown colour or precipitate with ammonia.

$$2K_2HgI_4 + 3NaOH + NH_4OH \xrightarrow{NH_3} \begin{array}{c} Hg \\ O \\ Hg \\ I \end{array} + 4KI + 3NaI + 3H_2O$$

Iodine of Millon's base

(c) When potassium cyanide is added to copper and cadmium salts solution, both copper and cadmium from complexes. These two complexes have different stabilities.

$$CuSO_4 + 2KCN \longrightarrow \underset{\text{ppt.}}{Cu(CN)_2} + K_2SO_4$$

$$2Cu(CN)_2 + 6KCN \longrightarrow \underset{\text{Pot. curprocyanide}}{2K_3[Cu(CN)_4]} + (CN)_2$$

$$CdSO_4 + 3KCN \longrightarrow Cd(CN)_2 + K_2SO_4$$

$$Cd(CN)_2 + 2KCN \longrightarrow \underset{\text{Pot cadmicyanide}}{K_2[Cd(CN)_4]}$$

When hydrogen sulphide is passed in these two solutions of complexes, cadmium is only precipitated because the copper complex is much more stable than the cadmium complex.

$$K_3Cu(CN)_4 \rightleftharpoons 3K+ + [Cu(CN)_4]^{3-}$$

$$[Cu^+ + (CN)_4]^{3-} \rightleftharpoons Cu^+ + 4CN^- \text{ (Negligible dissociation)}$$

$$K_2[Cd(CN)_4] \rightleftharpoons 2K^+ + [Cd(CN)_4]^{2-}$$

$$[Cd(CN)_4]^{2-} \rightleftharpoons Cd^{2+} + 4CN^- \text{ (High dissociation)}$$

(d) Separation of nickel and cobalt : When potassium cyanide is added to a mixture of cobalt and nickel salts, potassium cobalto cyanide is formed whereas nickel only forms nickel cyanide. Potassium cobalto cyanide with bromine and alkali is converted into potassium cobalticyanide whereas nickel cyanide with bromine and NaOH yields a black ppt. of nickel oxide.

$$CoCl_2 + 2KCN \longrightarrow Co(CN)_2 + 2KCl$$

$$Co(CN)_2 + 4KCN \longrightarrow K_4[Co(CN)_6]$$

$$2K_4[Co(CN)_6] + H_2O + O \longrightarrow 2K_3[Co(CN)_6] + 2KOH$$

$$NiCl_2 + 2KCN \longrightarrow Ni(CN)_2 + 2KCl$$

(e) Reaction with dimethyl glyoxime : Ni^{2+} form a red ppt. with dimethyl glyoxime. This reaction has been used for the separation of nickel from cobalt in the group IV.

$$2\begin{array}{l} CH_3 - C = N - OH \\ \quad\quad | \\ CH_3 - C = N - OH \end{array} + 2NH_4OH + NiCl_2 \longrightarrow$$

Dimethy lg lyoxime

(f) Reactions with yellow ammonium sulphide : The reaction of yellow ammonium sulphide with sulphides of IInd group cations is used for the separation of IIA from IIB. IIB cations form soluble complexes while of IIA cations do not form.

Arsenic,

$$As_2S_3 + 3(NH_4)_2S \longrightarrow \underset{\text{soluble}}{2(NH_4)_3AsS_3}$$

$$As_2S_5 + 3(NH_4)_2S \longrightarrow \underset{\text{soluble}}{2(NH_4)_3AsS_4}$$

Antimony,

$$Sb_2S_3 + (3NH_4)_2S \longrightarrow \underset{\text{soluble}}{2(NH_4)_3SsS_3}$$

$$Sb_2S_5 + 3(NH_4)_2S \longrightarrow \underset{\text{soluble}}{2(NH_4)_3SbS_4}$$

Tin

$$SnS + (NH_4)_2S \longrightarrow \underset{\text{soluble}}{(NH_4)_2SnS_2}$$

$$SnS_2 + (NH_4)_2S \longrightarrow \underset{\text{soluble}}{(NH_4)_2SnS_2}$$

These thioarsenite, thioarsenate, thioantimonite, thioantimonite, thiostannanite and thiostannate are soluble complexes from which original sulphides of arsenic, antimony and tin can be ppted by making the solution acidic with dilute hydrochloric acid.

(f) Complexes are usually formed to keep a metal ion in solution or to dissolve precipitate and bring the metal ion back into solution. For example, AgCl dissolves on addition of ammonia solution and forms the complex ion, diamine silver (I) ion.

$$AgCl + 2NH_3 \longrightarrow [Ag(NH_3)_2]^+ + Cl^-$$

(g) Formation of complexes with characteristically coloured solutions or precipitates is used for the detection of metal ion in solution. For example, Cu^{2+} ion can be detected in solution by the dark blue coloured solution it forms on addition of ammonia.

$$Cu^{2+} + 4NH_3 \longrightarrow \underset{\text{dark blue}}{[Cu(NH_3)_4]^{2+}}$$

(h) Use in gravimetric determination : Inner complexes are often insoluble in aqueous medium but soluble in organic solvents. The formation of such chelates often needs suitable pH range and many metal ions can be quantitatively precipitated and metal ions determined gravimetrically.

Some applications include estimation of aluminium as yellow coloured tris (8–hydroxy quinolonate), aluminium (III) and copper (II) as light green coloured bis-(quinolonate) aluminium (III) and copper (II). Some complexes are unstable at the drying temperature (120–150°C) and such complexes are ignited to metal oxides.

(i) Use of organic sequestering agents in removal of interference in gravimetric estimations :

An organic ligand may read with more than one metal ion to form sparingly soluble precipitates. In such cases direct estimation of a particular metal ion in the presence of interfering ions is not possible.

A strong sequestering agent like EDTA is of great importance in making interfering ions ineffective, and thus making the precipitating ligand almost specific for a particular metal ion. For example, in the presence of EDTA, beryllium may be precipitated with ammonia in presence of chromium, cobalt, cadmium, iron, copper, lead, manganese, zinc, aluminium, bismuth etc.

(j) Solvent extraction and separation of metal ions through complexation : Both iron (III) and aluminium (III) form complexes with 8-hydroxy quinoline. Iron (III) complex is easily soluble in chloroform. Iron (III) can be separated from aluminium (III) around pH 3.0 by the addition of 8–hydroxy–quinoline (oxine), followed by extraction with chloroform.

(k) Complexometric titrations : The sequestering effect of EDTA has led to the development of some versatile titrimetric reagents for the estimation of metal ions. Mostly EDTA titrations are performed with indicators, which are chelating agents whose metal complexes have different colours from the reagent.

The indicators most commonly employed are Erichrome black T, murexide, etc. Such complexometric titrations are used in the determination of water hardness. This procedure affords various advantages as compared to old and laborious palmitate titrations.

(I) Spectrophotometric estimation of micro amounts of metal ions. A large number of metal complexes are intensely coloured which provide a basis for their spectrophotometric determination when present in trace amounts.

For example, iron (II) forms dark red complex with o-phenanthroline over a wide range of pH 2–9. Trace amounts ($\sim 10^{-6}$ gm/ml.) of iron (II) can be estimated by measuring the optical density of the complex at 5.5 mμ.

3. Photography : Formation of $[Ag(S_2O_3)_2]^{3-}$ complex ion has an important role in photography. After the exposed film is developed, sodium thiosulphate solution is used to dissolve the unreduced silver halide.

$$AgBr + 2S_2O_3^{2-} \longrightarrow [Ag(S_2O_3)_2]^{3-} + Br^-$$

4. Extraction of Metals from Ores : The noble metals like Ag and Au are extratced from their ores through the formation of cyanide complexes like $[Ag(CN)_2]^-$ and $[AuCl_4]^-$ respectively.

5. Artificial Silk : Schweitzer's reagent, *i.e.*, tetrammine copper (II) hydroxide $[Cu(NH_3)_4](OH)_2$ is used as a solvent for cellulose during the manufacturing of artificial silk.

6. Biological Applications : A number of metal complexes are of biological importance. In particular (i) haemoglobin in the red blood cells contains an iron-porphyrin complex, (ii) chlorophyll in green plants contains magnesium-porphyrin complex, (iii) vitamin B^{12} is a cobalt complex.

Body contains a number of compounds like adrenaline, citric acid and cortisone which form unwanted complexes with metals like lead, copper etc. and prevent normal metabolism. Lead poisoning and copper poisoning are treated by injecting EDTA, so that metal–EDTA complex is excreted in the urine.

COMPARISON OF VALENCE BOND THEORY AND CRYSTAL FIELD THEORY

Some of the properties of complexes which could not be explained on the basis of valence bond theory are satisfactorily explained by crystal field theory. CFT is thus definitely an improvement over VBT. The following merits of CFT over VBT will prove that statement :

(i) CFT predicts a gradual change in magnetic properties of complexes rather than the abrupt change predicted by VBT.

(ii) In some complexes, when Δ_0 is very close to P, simple temperature changes may affect the magnetic properties of complexes. Thus, CFT provides theoretical basis for understanding and predicting the variation of magnetic moments with temperature as well as the detailed magnetic properties of complexes. This is just in contrast to VBT which cannot predict or explain magnetic behaviour beyond the level of specifying the number of unpaired electrons.

(iii) Though the assumptions inherent in VBT and CFT are vastly different, the main difference lies in their description of the orbitals not occupied in the low spin states. VBT forbids their use as they are involved in forming hybrid orbitals, while CFT strongly discourages their use as they are repelled by the ligands.

(iv) According to VBT, the bond between the metal and the ligand is covalent, while according to CFT it is purely ionic. The bond is now considered to have both ionic and covalent character. Unlike valence bond theory.

(v) CFT provides a framework for the ready interpretation of such phenomena as tetragonal distortions.

(vi) CFT provides satisfactory explanation for the colour of transition metal complexes, *i.e.*, spectral properties of complexes.

(vii) CFT can semiquantitatively explain certain thermodynamic and kinetic properties.

(viii) CFT makes possible a clear understanding of stereochemical properties of complexes.

Ligand Field Theory

Recently based on crystal field theory of *Bethe* (1929) and *Van Vleck* (1931–1935), a new approach known as ligand field theory was postulated. In the crystal field theory, attraction between central metal and ligand in a complex is regarded as purely elctrostatic, *i.e.,* the bonding between the central metal ion and ligands is purely ionic.

Thus, in crystal field theory covalent character of bond between metal and ligand is not taken into account. However, the enough evidence is available which suggests there is some measure of covalent bonding in complexes. The various experimental evidences are as follows.

(i) *Most direct evidence is obtained from electron* ***resonance spectrum*** *(ESR) of complexes. ESR spectrum of* $[IrCl_6]^{2-}$ *clearly shows hyperfine splitting indicating the delocalization of d electrons onto six chlorines. Such study of other complexes too gives similar results.*

(ii) ***Nuclear magnetic resonance (NMR)*** *studies of the complexes like* $KMnF_3$ *and* $KNiF_3$, *definitely indicate that the metal* t_{2g} *and* e_g *electrons pass a fraction of time around the fluorine nuclei.*

(iii) ***Nuclear quadrupole resonance (NQR)*** *spectra of square planar halide complexes of divalent platinum and palladium suggest considerable covalency in the metal-ligand bonds.*

(iv) ***The unusually large absorption band intensities*** *observed for tetrahedral complexes like* $[CoCl_4]^{2-}$ *have been interpreted as due to appreciable covalent nature of the metal-ligand bonds.*

The ligand field theory is eventually the same as pure crystal field theory but covalent character being taken into account.

When the orbitals overlap, *i.e.,* covalent character is excessive as in metal complexes of carbon monoxide or the isocyanide, the *molecular orbital theory* gives a more complete explanation of the *metal-ligand bonding.*

Ligand field theory differs from the valence bond theory by recognising the existence of antibonding as well as the bonding and non-bonding orbitals. For the present, the ligand field theory seems to offer the most practical approach to bonding in co-ordination compounds.

In ligand field theory, it is assumed that molecular orbitals are formed by the overlap of orbitals from the ligand with the atomic orbitals of the central atom. (The procedure is similar as adopted in case of simple molecules). The molecular orbitals thus formed may be of a *bonding, antibonding or a non-bonding character.* The antibonding orbitals are similar to the bonding orbitals except that the antibonding orbitals lie higher in energy and have nodes or regions of low electron density between the central atom and the ligands. The antibonding orbitals are of interest because it is these into which electrons may be excited from the *tis* orbitals by absorptions of energy. The non-bonding orbitals in octahedral complexes are simply the *dxy, dxz* and *dyz* orbitals of the central atom.

As the number of molecular orbitals formed is always equal to the number of atomic orbitals taking part in the overlappings and this number is quite large for complexation processes. The MO energy level diagrams for metal complexes are highly complicated. The fact that σ as well as π (both types of overlaps are involved in the coordination process) add more complication to these diagrams.

The three common types of complexes are octahedral, tetrahedral and square planar. But we shall discuss octahedral complexes.

THE WATER IONS SYSTEM

According to Arrhenius theory of ionisation, all substances which give hydrogen ions when dissolved in water could be called acids while, those which ionise in water to form hydroxyl ions, coula be called bases.

Examples of Arrhenius acids are HCl, H_2SO_4, CH_3COOH, etc. Examples of Arrhenius bases are NaOH, NH_4OH, etc.

Acid-base equilibria could now be treated in a quantitative way and acid-base neutralisation could he considered to result from the union of the hydrogen ions with hydroxyl ions to form water molecules,

$$H^+ + OH^- \longrightarrow H_2O$$

When we write a general equation.

$$HA \longrightarrow H^+ + A^-$$

for an acid, suggesting the formation of a group, H^+, but this is not very correct as free protons could exist only when gaseous hydrogen atoms are ionised. In the presence of protophilic liquids such as water, unattached proton could not exist; in water a proton gets attached to a number of water molecules by co-ordinate linkages, releasing a very large amount of energy. The proton in its hydrated state is usually expressed by H_3O^+, even though the number of water molecules attached to one proton is more than one. The hydrated proton is called the hydroxonium or hydronium ion. On the other hand hydroxyl ions are also hydrated in water through hydrogen bonds and are represented by OH- formula only because of convenience.

Successes

(1) Arrhenius concept has been invaluable in elucidating the behaviour of aqueous solutions. The constant heat of neutralisation of a strong acid by a strong base is readily explained in the light of this concept.

(ii) This theory correlated the catalytic action of acids with the concentration of hydrogen ions.

(iii) This concept also lends to quantitative determinations of acid or base strengths front the determination of following equilibrium relation:

$$K = \frac{{}^{a}H^{+} \, {}^{a}H^{-}}{{}^{a}HB}$$

Limitations

(i) The Arrhenius concept lacks in generality.

(ii) The most serious limitation is that it deals with dissociation and acid-base reactions in aqueous medium only and does not mention the role of non-aqueous solvents in the dissociation of acids and bases into ions. According to this concept, HCl is regarded as an acid only when dissolved in H_2O and not in some other solvents like benzene, chloroform, etc. or when it exists in the gaseous form.

(iii) This concept restricts bases merely to hydroxides. Thus, even metal oxides would not be regarded as bases.

(iv) According to this concept, a neutralisation process is restricted to aqueous medium alone but many non-aqueous solvents like liquid NH_3 liquid SO_2, etc. are known to undergo comparable acid-base neutralisation reactions.

(v) It cannot explain the acidic behaviour of certain salts such as $AlCl_3$ in water.

(vi) This concept does not elaborate the fate of dissociated hydrogen' ions in solution because of their high hydrational energy (1075 kj/mole).

(vii) A proton (H^+) is non-existent and is generally present as the hydrated H_3O^+ ion or H_9O_{4+} species.

(viii) A proton is also formed in other solvents like NH_3, ROH, etc.

(ix) There are nonprotonic and nonhydroxylic compounds which may yield H^+ or OH ions in water.

$$SO_3 + H_2O \rightarrow H_2SO_4 \rightleftarrows H^+ + HSO_4^-$$

$$CaO + H_2O \rightarrow Ca(OH)_2 \rightleftarrows Ca^{2+} + 2OH^-$$

THE PROTON-DONORACCEPTORSYSTEM

Bronsted and Lowry in 1923 proposed a new concept to acids and bases which is independent of solvents. According to atom.

"An acid is any hydrogen contained substance (a molecule or a cation or an anion) whereas a base is any substance (a molecule or cation or an anion) that can accept a proton from any other substance."

From the above definition, it follows that an acid is a proton donor and base is a proton acceptor. It also follows from this definition that it is not essential for bases to have OH^- groups as demanded by the Arrhenius theory.

A neutralisation process, therefore, involves a release of a proton by the acid, and the acceptance of that proton by a base. This concept makes acid-base reactions also to be known as protolytic reactions independent of solvent.

$$\underset{\text{acid}}{H} : Cl + \underset{\text{base}}{NH_3} \rightarrow Cl^- + NH_4^+$$

$$\underset{\text{acid}}{NH_4^+} + \underset{\text{base}}{OH^-} \rightarrow NH_3 + H_2O$$

Bronsted and Lowry acids and bases are of the following types :

Bronsted-Lowry Acids

$$\begin{bmatrix} \text{Molecular : } HCl \rightarrow H^+ + Cl \\ \text{Cationic : } [A(H_2O)_6]^{3+} \rightarrow H^+ + [A(H_2O)_5OH]^{2+} \\ \text{Anionic : } H_2PO_4^- \rightarrow H^+ + HPO_4^{2-} \end{bmatrix}$$

Bronsted-Lowry Bases

Molecular : $Py + H^+ \rightarrow]PyH]^+$; where Py = Pyridine
Cationic : $[A(H_2O)_5OH]^{2+} + H^+ \rightarrow [A(H_2O)_6]^{3+}$
Anionic : $PO_4^{3-} + H^+ \rightarrow HPO_4^{2-}$

Substances like H_2O and HCO_3 ions who donate as well as accept protons.

$$H_2O \rightleftarrows H^+ + OH^- \text{ (donation)}$$
$$H_2O + H^+ \rightleftarrows H_3O^+ \text{ (acceptance)}$$
$$HCO_3^- \rightleftarrows H^+ + CO3 \text{ (donation)}$$
$$HCO_3^- + H^+ \rightleftarrows H_2CO_3 \text{ (acceptance)}$$

are classed as amphoteric substances

The Bronsted-Lowry concept is quite useful For example:

(i) This concept also takes into account protonic solvents like NH_3, H_2SO_4, etc.

$$NH_3 + NH_3 \rightleftarrows NH_4^+ + NH_2^-$$
$$H_2SO_4 + H_2SO_4 \rightleftarrows H_3SO_4 + HSO_4$$

(ii) All-proton transfer reactions may be treated as follows:

$$NH_4^+ + S^{2-} \rightarrow NH_3 + HS$$

Conjugate Acids and Bases

A study of the following reactions for processes occurring in water brings out another important feature of this concept about acids and bases:

$$CH_3COOH + OH^- \rightleftarrows HOH + CH_3COO^-$$
$$H_2O + H_2O \rightleftarrows H_3O^+ + OH^-$$
$$NH_4^+ + CO_3^{2-} \rightleftarrows HCO_3^- + NH_3$$
$$H_2O + CO_3^{2-} \rightleftarrows HCO_3^- + OH^-$$

From the above reactions, it is seen that when a molecule or an ion loses a proton, it is necessary to convert into a form which would accept a proton, to reform the original acid. The latter form is therefore, a base and it is called a linked or conjugate base to the initial acid. Such pairs of substances which are formed from one another by the gain or loss of a proton are called conjugate acid-base pairs. Acetic acid and acetate ion, H_2O and OH^- ion and NH_4^+ ion and NH_3 , H_3O^+ ion and H_2O, HCO_4 ion and CO_3^{2-} ion are all examples of conjugate pairs of acids and bases.

Just as for every weak acid we have a conjugate base, for every weak base we have a conjugate acid which is formed by the addition of a proton to the base. The strong acids and bases are not covered by this statement for the lack of the corresponding proton combination.

Furthermore, a substance like HNO_3 behaving as an acid or a base depends on the solvent uses. In liquid NH_3, H_2O and even dry CH_3COOH, HNO_3 acts like an acid by donating protons, but in dry HF medium, it accepts protons, and behaves like a base.

Relative Strength of Acids and Bases

According to Bronsted concept, the strength of an acid is regarded as a measure of its tendency to donate proton. Thus, a strong acid is that one which has a strong tendency to donate a proton. Similarly, a strong bases is that one which has a strong tendency to accept a proton.

Generally, two methods are employed to compare the relative strength of given acids. These are as follows

(i) In the first method we compare the dissociation constants or ionisation constants of acids/bases to get idea about their strength. Higher the dissociation constant, greater the strength of acid/base. For example, we are interested in comparing the acidic strength of CH_3COOH and HCN. The values of dissociation constant of CH_3COOH and HCN are 1.8×10^{-5} and 4.0×10^{-10} respectively.

$$CH_3COOH + H_2O \rightleftarrows H_3O+ + CH_3COO^-(K_a = 1.8 \times 10^{-5})$$

$$HCN + H_2O \rightleftarrows H_3O^+ + CN^- \quad (Ka = 4.0 \times 13^{-10})$$

As the value of K_a of CH_3COOH is greater than that of HCN, it means that CH_3COOH is a stronger acid than HCN and CN^- ion is a stronger base than CH_3COO^- ion

(ii) The second method is known as competitive protolysis method. In this method, one acid is added to the conjugate base of another acid, and, then, one has to determine the equilibrium concentrations experimentally.

For example, if $NaOC_2H_5$ is added to H_2O, it is exprimentally found that $C_2H_5O^-$ has good tendency to react completely with the H_2O to from C_2H_5OH and OH^- ion. It means that $C_2H_5O^-$ is a stronger base than OH^- and H_2O is a stronger acid than C_2H_5OH.

$$\underset{\text{Stronger acid}}{H_2O} + \underset{\text{Stronger base}}{C_2H_5O^-} \rightleftarrows \underset{\text{Weaker acid}}{C_2H_5OH} + \underset{\text{Weaker base}}{OH^-}$$

It is important to mention here that acidity increases with the increase of positive charge whereas the basicity increases with the increase of positive charge. Thus, $[Fe(H_2O)_6]^{3+}$ is a stronger acid than $[Fe(H_2O)_6]^{2+}$ ion and $[Ni(OH)_4]^{2-}$ is a stronger base than $[Ni(OH)_4]^-$ ion.

Successes

(i) This concept shows that acid base phenomena are dependent on the solvent.

(ii) The concept covers a large number and variety of substances under base, for example, hydroxide ion, amide ion, ethoxide ion, pyridine, alcohols, acetate ion, bisulphate ion and many others.

(iii) Bronsted-Lowry concept is useful in accounting for the hydrolysis of salt solution. A solution of $FeCl_3$ is acidic because the proton donor ability of hydrated ferric ion exceeds the proton acceptor ability of the chloride ion and measurable excess of $[Fe(H_2O)_x]^{3+}$, ion appears in the solutions.

$$FeCl_3 + {}_xH_2O \rightleftarrows [Fe(H_2O)_x]^{3+} + 3Cl^-$$

Limitations

This theory serves excellently in protonic solvents but fails in case of some obvious acid-base reaction. For example, the formation of H_2SO_4 by the reaction of SO_3 and H_2O, the acidic participant SO_3 in this case has no protons to give.

LEVELLING SOLVENTS

The apparent strength of a protonic acid has been found to depend upon the solvent in which the acid is dissolved. When all the acids which are stronger than H_3O^+ (*e.g.*, $HClO_4$, H_2SO_4, HCl and HNO_3) are added to H_2O, they lose a proton to H_2O forming H_3O^+ ion. All these acids have equal strength because all these acids have been levelled to the strength of H_3O^+ ion which remains in solution and is common to all such solutions. This phenomenon in which the strength of all the acids becomes equal to that of H_3O^+ ion is known as levelling effect of the solvent *i.e.*, water. Here water is known as a levelling solvent for all these acids.

Solvent	*Acid*	+ *Base*	→ *Salt*	+ *Solvent*
H_2O	H_3O^+ (HBr)	OH^- (KOH)	K^+, Br^-	$2H_2O$
C_2H_5OH	$C_2H_5OH_{2+}$ (HBr)	$O^-C_2H_5$ (KOC_2H_5)	K^+, Br^-	$2C_2H_5OH$

The solvents in which complete transfer takes place are known as levelling solvents. It means that the solvents in which the solute is 100% ionised are known as levelling solvents. For example, liquid NH3 acts as a levelling solvent for HF and HCI because both these are -100% ionised in liquid NH_3 to give-100% NH_4^+ ions.

In H_2O, HF undergoes partial ionisation while HCI and HBr undergo –100% ionisation. Thus, H_2O is a differentiating solvent for HF but for HCl and HBr it is a levelling solvent.

Other solvents may have their own levelling effects, too. Thus, benzoic acid is a very weak acid in aqueous solution. But in liquid ammonia it acts as a very strong acid. In general, it may be expected that acidic solvents (like acetic acid) will exert less levelling effect on acid strengths (and similarly-basic solvents on base strengths). Relative strengths of the common mineral acids have been obtained by conductace measurement in anhydrous acetic acid. The order found is

$$HClO_4 > HBr > H_2SO_4 > HCl > HNO_3$$

SOLVENT SYSTEM CONCEPT

Franklin extended the dissociation principles to non-aqueous solvents and realised that non-aqueous solvent molecules may also dissociate into two oppositely charged ions. According to him, if an acid releases H^+ in water to form H_3O^+ ions then an ammonium salt, NH_4Cl, by releasing NH_4^+ in liquid ammonia, must also be termed an acid. If NaOH, by virtue of its releasing OH^- ion in aqueous medium, is termed a base in that medium, then $NaNH_2$ should also be termed a base in liquid NH_3 since it releases the amide, NH_2^- ion. The dissociation (or auto-ionisation) of non-aqueous solvent is responsible in such solvents.

$$NH_3 \rightleftarrows H^+ + NH_2^-$$

According to Franklin, an acid is a substance which by dissolution in the solvent forms the same cation as does the solvent itself due to autoionisation. A base is one that gives on dissociation in the solvent the same anion as does the solvent itself on its own ionisation.

Solvent	*Acid +*	*Base →*	*Salt +*	*Sovent*
H_2O	H_3O^+ (HBr)	OH^- (KOH)	K^+, Br^-	$2H_2O$
C_2H_5OH	$C_2H_5OH_2^+$ (HBr)	$O^-C_2H_5$ (KOC_2H_5)	K^+, Br^-	$2C_2H_5OH$

The solvent system concept to non-protonic solvent *e.g.*, SO_2, $COCl_2$ in case of acids and bases was applied by Cady and Elsey.

According to them, substances which increase the concentration of the cation characteristics of the solvent are acids, whereas substances which increase the concentration of the anion characteristics of the solvents are bases.

In N_2O_4 as a solvent, substances such as NOCl which yield NO^+ ions behave as acids, and substances such as $NaNO_3$ which yield NO_3^- ions behave as bases.

The auto-ionisation of some non-protonic solvents is given below:

$$\begin{array}{ccc} \text{Acid Base} & & \text{Acid Base} \\ SO_2 + SO_2 & \rightleftarrows & SO^+ + SO_3 \\ COCl_2 + COCl_2 & \rightleftarrows & COClCOC_2^+ + Cl^- \\ N_2O_4 + N_2O_4 & \rightleftarrows & 2NO^- + 2NO_3^+ \\ BrF_3 + BrF_3 & \rightleftarrows & BrF_2 + BrF_4^- \end{array}$$

Just as with Arrhenius concept neutralisation is reaction between an acid and a base to produce a salt and the solvent. Neutralisation reactions in non-aqueous media are given below:

Acid Base Salt Solvent

In liquid NH_3: $NH_4Cl + NaNH_2 \rightleftarrows NaCl + 2NH_3$

In liquid SO_2: $SOCl_2 + Na_2SO_3 \rightleftarrows 2NaCI + 2SO_2$

It is clear from the following reactions that there is complete analogy between solvolytic and amphoteric behaviours in aqueous and non-aqueous solvents.

Solvolytic Behaviour

$$\left[\begin{array}{ll}\text{In liq. } NH_3 & : AlCl_3 + NH_3 \rightarrow [Al(NH_2)]^{2+} + H^+ + 3Cl^- \\ \text{In } H_2O & : AlCl_3 + H_2O \rightarrow [Al(OH)^{2+} + H^+ + 3Cl^-\end{array}\right.$$

Amphoteric behaviour:

$$\left[\begin{array}{l}\text{In liq } NH_3: Zn(NH_2)_2 + 2NH_2^- \rightarrow [Zn(NH_2)_4]^{2-} \\ \text{In } H_2O : Zn(OH)_2 + 2OH^- \rightarrow [Zn(OH)_4]^2\end{array}\right.$$

Success

This general theory of solvent system concept possesses the obvious advantage of relating acid-base behaviour in aqueous and non-aqueous solvents, both protonic and nonprotonic.

Limitations

(i) It limits acid-base phenomena to solvent systems only.

(ii) The concept does not explain the neutralisation occurring without the absence of solvent.

(iii) It does not explain the acid-base processes which may occur in the absence of solvent.

(iv) In low dielectric constant solvents, the existence of ions is not attainable.

(v) Emphasis has been laid mainly on the chemical properties of the solvent. However, the physical properties have been completely ignored.

(vi) Too much stress has been laid mainly on the ionic reaction which actually even dielectric affects would be able to explain the reactions easily. For instance, Gutman and his coworkers (1952-60) have explained reactions between the chloride ion acceptors ($FeCl_3$,$SnCl_4$, etc.) and chloride ion donors (Me_4NCl) in $OPCl_3$ as follows:

$$OPCl_3 \rightleftarrows OPCl_2^+ + Cl^- \text{ (autoionisation)}$$

$$Me_4NCl \rightleftarrows Me_4N^+ + Cl^- \text{ (base)}$$

$$OPCl_3 + FeCl_3 \rightleftarrows OPCl_2^+ + FeCl_4^- \text{ (acid)}$$

However, Meek and Drago (1962) showed that the reaction can be carried out equally well in triethyl phosphate medium without involving chloride ions.

It means that the dielectric constant of the medium is more important than the ionisation reaction of the solvent. Thus the solvent system concept for acid and bases simply extends, but does not necessarily improve, the Arrthenius waterions system.

LUX-FLOOD CONCEPT

Lux proposed a theory to explain the acid-bases concept of oxide system. According to the concept, a base is any material which gives up oxide ions and an acid is any material which gains oxide ions. Some typical reactions are

$$\text{Base} \rightleftarrows \text{Acid} + {}_xO^{2-}$$

$$SO_4^{2-} \rightleftarrows SO_3 + O^{2-}$$

$$CaO \rightleftarrows Ca^{2+} + O^{2-}$$

Hence according to the concept, a bases is an oxide donor and acid is an oxide ion acceptor. This view is particularly useful to high temperature chemistry, as in the fields of ceramics and metallurgy.

For example, the following reactions

$$CaO + SiO_2 \rightarrow CaSiO_3$$

$$PbO + SO_3 \rightarrow PbSO_4$$

involving basic oxides (CaO,PbO) and acidic oxides (SO_3, SiO_2) are reacting to form salts.

Furthermore, according to this theory amphoteric substances are those which show both a tendency to take up or give up oxide ions depending upon the circumstances, *i.e.*,

$$Na_2O + \underset{\text{Acid}}{ZnO} \rightarrow 2Na^+ + ZnO^2$$

The oxide transfer picture due to Lux can be extended to any negative ion, *i.e.*, halide, sulphide, etc.

$$3NaF + AlF_3 \xrightarrow[\text{temperature}]{\text{high}} 3Na^+ + AlF_6^{2-}$$

The Lux-flood definition can be extended to include transfer of any anion *e.g.*, halide, sulphide, etc.

Limitations

The limitations of such a concept are obvious. This view can be reduced to one part of the more general theories of Lewis and Usanovich.

LADY-ELSEY CONCEPT

Bases on the characteristic cation or anion of the solvent which is produced by it with the solute, a more general definition of acids and bases applicable to all solvents was provided by Cady and Elsey. Consider, for example.

(i) $$H_2O + H_2O \rightarrow H_3O^+ + OH^-$$

(ii) $$NH_3 + NH_3 \rightarrow NH_4^+ + NH_2^-$$

(iii) $$HF + HF \rightarrow H_2F^+ + F^-$$

(iv) $$N_2O_4 + N_2O_4 \rightarrow 2NO^+ + 2NO_3^-$$

(v) $$CH_3COOH + CH_3COOH \rightarrow CH_2COOH_2^+ + CH_3COO^-$$

The characteristic cations furnished by water, ammonia, hydrogen fluoride and nitrogen tetroxide from (i), (ii), (iii) and (iv) above are respectively H_3O^+, NH_4^+, H_2F^+ and NO^+. The respective characteristic anions of these solvents are OH^-, NH_2^-, F^- and NO_3^-.

According to Cady-Elsey Concept

An acid is a substance which in dissolution in solvent gives the cations characteristic of the solvent and a base is a substance which yields anion characteristic of the solvent either by direct dissolution or by reaction with the solvent.

In the light of this concept, it is seen from (i), (ii), (iii) and (iv) above that: All substances which furnish H_3O^+ ions in water solvent, NH_4^+ ions in ammonia solvent, H_2F^+ in hydrogen fluoride solvent, NO^+ in nitrogen tetroxide and acetic acid medium are termed as bases. For example,

$$NH_4X \rightleftarrows NH_4 + X^-$$

in ammonia is an acid and $MNH_2 \rightarrow M^+ \; NH_2^-$ in it is ao base. BF_3 in hydrogen fluoride reacts as $BF_3 + 2HF \rightleftarrows H_2F^+ + BF_4$ forming H_2F^+ ions. It is therefore an acid.

ELECTRON-DONOR-ACCEPTOR SYSTEM OR THE LEWIS CONCEPT

Gilbert N. Lewis proposed a broader concept of acids and bases that liberated acid-base phenomena from the proton; although Lewis first proposed his system in 1923, he did little to develop it until 1938, Lewis defined a base as a substance that has an unshared electron pair with which it can form a covalent bond with an atom, molecule, or ion.

An acid is a substance that can form a covalent bond by accepting an electron pair from a base. An acid must have a vacant orbital into which an electron pair donated by a base can be accommodated.

An example of a Lewis acid-base reaction that is not treated as such by any other acid-base concept

```
        :F̈:      H                      :F̈:      H
         |       |                       |       |
  :F̈  –  B  +  :N  –  H   ⟶   :F̈  –  B  ←  N  –  H
         |       |                       |       |
        :F:      H                      :F:      H
         ̈                               ̈
Acid (acceptor)  Base (donor)                Complex
(Electrophile)   (Nucleophile)
```

Fig. 1

Lewis acid is often described as acceptor or electrophile, and the Lewis base is called a donor and nucleophile. On the basis of this concept, cations are acid-base complexes.

Classification of Lewis acids and bases : Based on the Lewis system are essentially the same as in Bronsted-Lowry system, since any molecule or ion which accepts protons does so because of the presence of an unshared pair of electrons. For example, OH^-, CF^-. NH_3, H^3O^+, CN^- etc., are bases on the Lewis as well on Bronsted-Lowry system.

The properties of an acid might be due to the availability of an émpty orbital for the acceptance of a pair of electrons.

Hence Lewis definition does not attribute acidity to any particular element but rather, to a unique electronic arrangement. According to this concept, acids are classified as

1. **Molecules possessing a central atom with less than an octet of electrons :** Lewis acids of this class are the electron deficient molecules such as B in BF_3, Al in $AlCl_3$.

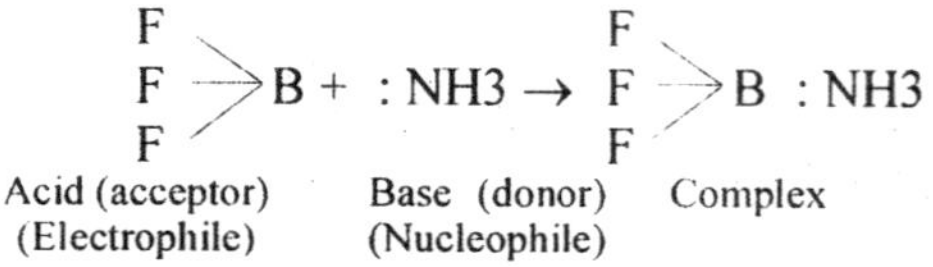

$$\begin{matrix} F \\ F \\ F \end{matrix} > B + :NH3 \rightarrow \begin{matrix} F \\ F \\ F \end{matrix} > B : NH3$$

Acid (acceptor) (Electrophile) — Base (donor) (Nucleophile) — Complex

Fig. 2

2. **Molecules in which the central atom has available 'd' orbitals and may acquire more than an octet of valence electrons:** Silicon tetrafluoride and tin (IV) chloride are typical examples:

$$\underset{\text{Lewis acid}}{SiF_4} + \underset{\text{Lewis base}}{2:\ddot{\underset{..}{F}}^-:} \rightarrow \underset{\text{Complex}}{SiF_6^{2-}}; \underset{\text{Lewis acid}}{SnCl_4} + \underset{\text{Lewis base}}{2:\ddot{\underset{..}{Cl}}^-;} \rightarrow \underset{\text{Complex}}{SCl_6^{2-}}$$

Some other examples of this type are

PF_3, PF_5, $TeCl_4$, SF_4, SeF_4, TiX_4, GeX_4,

These halides tend to form adducts with a number of organic bases.

3. **Cations :** This class of Lewis acids is made up of positively charged heavy metal ions with incomplete stable orbitals. Theoretically, all cations are potential Lewis acids because they are electron deficient.

   However, this property is negligible for the alkalimetal cations such as Na^+, and K^+, and is weak for the alkaline earth cations such as Ca^{2+}, Sr^{2+} and Ba^{2+}.

   Of the alkaline earth group, only Mg^{2+} and Be^{2+} show an appreciable tendency to behave as Lewis acid. The reaction of M^{3+} type of ion with water molecules is an example of Lewis acid-base interaction.

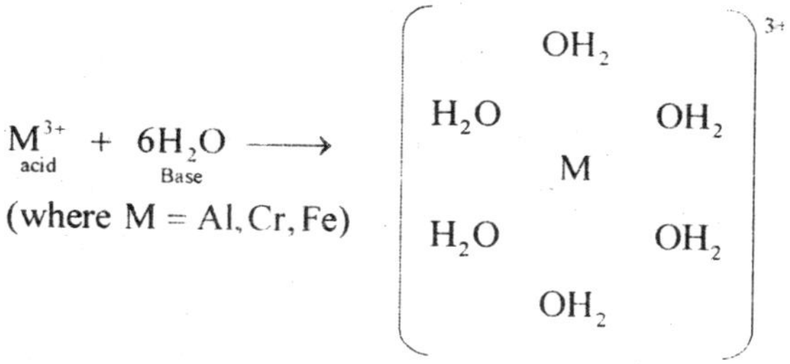

$$\underset{\text{acid}}{M^{3+}} + \underset{\text{Base}}{6H_2O} \longrightarrow \begin{bmatrix} & OH_2 & \\ H_2O & & OH_2 \\ & M & \\ H_2O & & OH_2 \\ & OH_2 & \end{bmatrix}^{3+}$$

(where M = Al, Cr, Fe)

Complex

Fig. 3

The reaction of Ag^+ with NH_3 is also typical

$$\underset{\substack{\text{Lewis}\\\text{acid}}}{Ag^+} + 2:\underset{\substack{\text{Lewis}\\\text{base}}}{NH_3} \rightarrow \underset{\text{Complex}}{[H_3N:\ Ag\ :NH_3]^+}$$

Usually, the number of reacting nucleophiles is twice the charge on the cation.

4. **Molecules with a multiple bond between atoms of dissimilar electronegativities :** An example of such a molecule is carbon dioxide, O = C = O. The oxygen atoms are more electronegative than the carbon atom. As a result, the electron density due to the it electrons is displaced away from the carbon atom, towards the oxygen atoms. The carbon atom is electron deficient and is apt to from a bond with a Lewis base such as OH^-

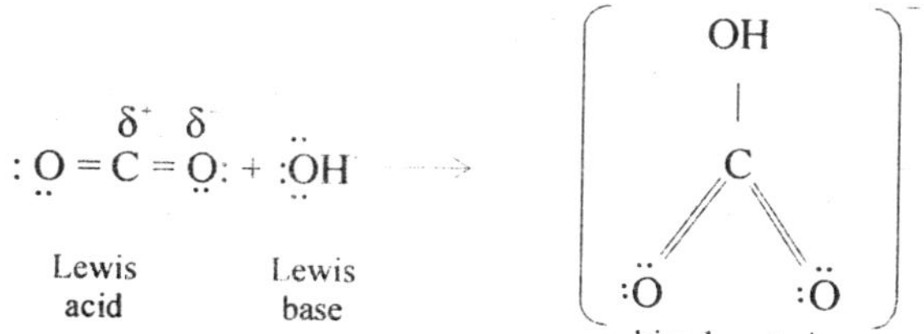

$$\underset{\text{Lewis acid}}{:\ddot{O} = \overset{\delta^+}{C} = \overset{\delta^-}{\ddot{O}}:} + \underset{\text{Lewis base}}{:\ddot{O}H^-} \longrightarrow$$

[bicarbonate structure: C bonded to OH, and to two :Ö atoms]$^-$

bicarbonate ion

SO_2 reacts like CO_2 towards OH^-,

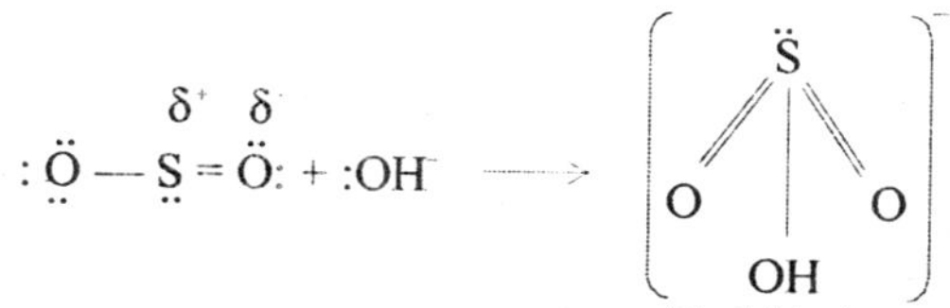

$$:\ddot{O} - \overset{\delta^+}{\ddot{S}} = \overset{\delta^-}{\ddot{O}}: + :OH^- \longrightarrow$$

[bisulphite structure: S̈ bonded to O, O and OH]$^-$

Bisulphite ion

Fig. 4

5. **Elements with an electron sextet :** Oxygen and sulphur atoms contain six electrons in their valency shells and, therefore, act as Lewis acids. The oxidation of sulphite to SO_4^{2-} by oxygen, and to $S_2O_3^{2-}$ by sulphur involves acid-reactions.

$$:\ddot{\underset{..}{O}}-SO_3^{2-} \rightarrow (O \leftarrow SO_3)^2$$

$$:\ddot{\underset{..}{S}}-SO_3^{2-} \rightarrow (S \leftarrow SO_3)^2$$

Successes

(i) Lewis concept is a broader interpretation of acid- base behaviour.

(ii) This theory is not dependent on the presence of one particular element, upon any given combination of elements, upon the presence of ions, or upon the presence or absence of a solvent.

(iii) The Lewis approach is, however, of great value in case where protonic concept is not applicable.

(iv) Lewis theory is frequently employed to interpret reaction mechanisms.

Limitations

(i) In Lewis approach there is a lack of uniform scale of acid and base strength. The strength of acid and base in terms of Lewis concept is variable and dependent on the reaction considered. In this respect it is inferior to Arrhenius concept or the Lowry concept.

(ii) The concept is so broad that it is unsuited to general chemistry, in which the explanation of acid-base behaviour is most important of an acid.

USANOVICH CONCEPT (POSITIVE-NEGATIVE SYSTEM)

Usanovich proposed definitions of acid and bases which are most comprehensive of all acid-base theories given from time to time. According to him, an acid is any species capable of giving cations, combining with anions or electrons, or neutralising a base to yield salt. In a similar way he defined base as any species capable of giving up anions or electrons, combining with cations, or neutralising an acid to yield a salt. The Usanivich theory includes all precious acid-base definitions and in this process it has become too general. It also includes oxidation-reduction reaction as a special class of acid-base reactions.

Usanovich has placed considerable emphasis on the central atom in the compound while accounting for the acid-base character of a substance. By co-ordination unsaturation he means the ability of an atom to increase its covalence. The acid function is determined by the presence of co-ordinately unsaturated positive particles while basic function by the presence of similarly unsaturated negative particles.

For instance, in SO_2 the central atom can be considered to be sulphur, and it is co-ordinately unsaturated. Thus, it is capable of accepting an anion such O^{2-}, thereby acting as an acid. In general acidity is promoted by highly charged positive particles (Na_2O is highly basic while Al_2O_3 is less basic) and basicity by highly charged negative particle.

Acid Base Relationships on the Basis of Usanovich Concept

Acid +	Base →	Salt	Justification
Cl_2 +	2Na →	2NaCl	Na loses an electron, Cl, gains, an electron
SO_2 +	Na_2O →	Na_2SO_4	Na_2O yields O^{2-}
Sb_2S_5 +	$3(NH_4)_2S$ →	$2(NH_4)3SbS_4$	$(NH_4)_2$ S gives up S^{2-}, Sb_2S_5 gains S^{2-}

Success

Usanovich concept is useful in classifying all examples of acidity and basicity. It is most general for all other definitions of acid-base so far given.

Limitations

(i) Usanovich concept is of great value for specific problems, but for general considerations, it is undoubtedly too broad to be of wide applicability.

(ii) The theory virtually involves that all chemical reactions fall into an acid-base category, and one begins to wonder tit the purpose of using any name other than chemical reaction.

Other theories. Konopic (1949) proposed donor-acceptor acids and bases which amount to little more than that of Usanovich concept.

Bjerrum (1951) considered acid as a proton donor while a conventional Lewis acid he referred to as an antibase. Mulliken (1952) developed quantum mechanical treatment of acids and bases. But all these theories add nothing new to our knowledge.

Actually we may say that each approach is correct as far as it goes and no conflict exists around them. But one should know the main theme of each concept and should adopt his thinking to the particular problem in hand.

SOFT AND HARD ACIDS AND BASES (THE SHAB PRINCIPLE)

Based on the preferential bonding, Arland, Chatt and Davies (1958) categorised the metal ions into two classes

1. **Class (a) :** This includes ions of alkali metals, alkaline earth metals, lighter transition metals in higher oxidation states like Ti^{4+}, Cr^{4+}, Fe^{3+} and Co^{3+}, and the hydrogen ion. H^+ These ions have the following characteristic features
   (i) They have small size, high polarising power and high oxidation state.
   (ii) Their outer electrons or orbitals are not easily distorted
1. **Class (b) :** This includes ions of the heavier transition metals and those in lower oxidation states such as Cu^+, Ag^+, Hg_2^{2+}, Pd^{2+}, Pt^{2+}, etc. These ions have the following characteristic features
   (i) These are large sized.
   (ii) Their outer electrons or orbitals are easily distorted.

Depending upon their behaviour towards the metals of classes (a) and (b), the ligands have been categorised by Arland, Chatt and Davies into the following two classes:

2. **Class (a) :** This includes ligands which preferentially combine with metal ions of class (a) For example, the ligands NH_3, R_3N, H_2O and F^- ions have great tendency to co-ordinate with class (a) metal ions. The tendency of the complexation of the ligands with class (a) metal ions follows the order given below

$$F > Cl > Br > I$$

O > S > Se > Te

N > P > As > Sb

2. **Class (b) :** This includes ligands which preferentially combine with metal ions of class (b). For example, ligands such as R_3P (phosphines) and R_2S have great tendency to co-ordinate with class (b) metal ions. The tendency of the complexation of the ligands with class (b) metal ions follows the order given below:

$F < Cl < Br < I$

$O \leq S < Se = Te$

$N \leq P < As < Sb$

Pearson (1963) suggested that terms hard and soft can be used for class (a) and class (b) respectively. Thus, in his classification, metal ions of class (a) are called hard acids and ligands of class (a) are called hard bases. On the other hand, metal ions and ligands of class (b) are called soft acids and soft bases respectively.

Pearson's Principle : The typical Lewis acid-base reaction can be generalized as

$$\underset{\substack{\text{Lewis acid} \\ \text{(Acceptor)}}}{A} + \underset{\substack{\text{Lewis base} \\ \text{(Donor)}}}{:B} \longrightarrow \underset{\text{Complex}}{A:B}$$

To broaden the scope of the reactions we assume that the bond in AB can be ionic, polar, or non-polar. It would be very beneficial if one could predict the stability of the complex A:B. A concept designed to explain it is called the principle of Soft and Hard acids and bases and was proposed by Pearson (1963).

Soft bases have donor atoms which are easily polarised and which have low electronegativity. These criteria are interdepedent, although not absolutely equivalent. They each relate to the ease with which electrons can be distorted away from the donor site.

Hard bases have properties opposite to those of the soft bases : the donor atoms tend to retain its electrons. Thus, hard bases have donor atoms with low polarizabilities and high electronegativities. Table I shows some typical acids classified as hard, borderline and soft.

Table 7 shows some typical bases classified as hard, borderline, and soft. Within a group in the periodic table, softness increases with increasing size of the donor atom. Thus, of the halide ions., F^- is the hardest and I^- is the softest. Hard base is hard to oxidize and soft base is easy to oxidize.

Table 7 : Classification of Acids

Hard Acids	*Borderline Acids*	*Soft Acids*
$H^+, Li^+, Na^+, K^+,$	Fe^{2+}, Co^{2+}, Ni^+,	$Co(CN)_5^{2-}$, Pd^{2+}, Pt^{2+}, Pt^{4+}
(Rb^+, Cs^+)	Cu^{2+}, Zn^{2+}	
$Be^{2+}, Be(CH_3)_2, Mg^{3+},$	$Rh^{3+}, Ir^{3+}, Ru^{3+},$	Cu^+, Ag^+. Au^+, Cd^{2+}, Hg^{2+}
$Ca^{2+}, Sr^{2+}, Ba^{2+}$	$B(CH_3)_3, GaH_3$	Hg^{2+}, CH_3Hg^+
		$BH_3.Ga(CH_3)_3, GaCl_3$
$Sc^{3+}, La^{3+}Ce^{4+}, Gd^{3+}, Lu^{3+},$	R_3C^+, $C_6H_5^+$,	$GaBr_3, GaTl^+$, $Tl(CH_3)_3$,
Th^{4+}, U^{4+}. UO_2^{2+} ,Pu^{4+}	Sn^{2+}, Pb^{2+}	
Ti^{4+}, Zr^{4+}, Hf^{4+}, VO^{2+}	NO^+, Sb^{3+}, Bi^{3+}	CH_2,carbenes
Cr^{3+}, Cr^{6+}, MoO^{3+}, WO^{4+}	SO_2	π- acceptors : trinitrobenzene,
$Mn^{2+}, Mn^{7+}, Fe^{3+}, Co^{3+}$		chloranil, quinones,
		Tetracyanoe thylene, etc.
$BF_3, BCl_3, B(OR)_3, Al^{3+},$		$HO^+, RO^+, RS^+, RSe^+,$
$Al(CH_3)_3$. $AlCl_3, AlH_3,$		Te^{4+}, RTe^+
Ga^{3+}, In^{3+}		
CO_2, RCO^+, NC^+, Si^{4+}		Br_2, Br^+, I_2, I^+,ICN,etc.
$Sn^{4+}, CH_3Sn^{3+}, (CH_3), Sn^{2+}$		O, Cl, Br, I, N, RO^-, RO_2.
N^{3+}, RPO_6^+, $ROPO_2^+$, As^{3+}		
SO_3, RSO_2^+, $ROSO_2^+$		M^+ (metal atoms) and bulk
Cl^{3+}, Cl^{7+} I^{5+}, I^{7+}		metals
HX (hydrogen-bonding molecules)		

Soft acids are those which have acceptor atoms large in size, of low positive charge and containing unshared pair (if electrons (p or d electrons) in their valence shell, *i.e.*, they have high polaritability and low electronegativity and do not have a noble gas configuration. Hard acids are those in which acceptor atoms are small in size, of high electronegativity and low polarizability and a noble gas configuration.

Table 8 : Classification of Bases

Hard Bases	*Borderline Bases*	*Soft Bases*
NH_3, RNH_2, N_2H_4	$C_6H_5NH_2$, C_5H_5N,	H^-
	N_3^-, N_2	
H_2O, OH^-, O^{2-}, ROH, CO	NO_2^-, SO_3^{2-}	R^-, C_2H_4, C_6H_6, CN^-, RNC,
RO^-, R_2O		
CH_3COO^-, CO_3^{2-}, NO_2^-,	Br^-	SCN^-, R_3P, $(RO)_3P$, R_2As
PO_4^{2-}, SO_4^{2-}, NO_2^-		R_2S, RSH, RS^-, $S_2O_3^{2-}$ I^-
F^-, (Cl^-)		

Soft acids form stable complexes' with bases that are highly polarisable, and are good reducing agents and not necessarily good bases towards the proton. Hard acids. On the other hand, usually form stable complexes with bases that bind well with protons.

We now apply the idea of soft and hard acids and bases to the question of the stability of the complex A : B The complex AB is most stable when A and B are either both soft or both hard. The complex is least stable when one of the reactants is very hard and the other is very soft

Pearson's principle is only an approximate qualitative prediction of the relative stability for the adducts and is not a theory or any explanation of the observations.

Theories of Hardness and Softness : Several theories have been put forward to explain the fact that the co-ordination compounds of soft acids and soft bases are most stable. The various theories are:

1. **Electronic theory :** According to the theory, hard-hard interactions involve ionic bonding and soft-soft interactions involve covalent bonding. It is expected that small sized and highly charged positive metal ions would he favouring ionic bonding with hard bases of class (a). If hard-soft interaction takes place, the resulting species would be unstable.

Misono and Coworkers (1967) proposed the following equation which correlates hardness and softness.

$$pK = -\log K = aX + bY + c$$

where K is the equilibrium constant for the dissociation of the metal-ligand complex, X,Y are the parameters of the metal ions, a and b are parameters for the ligands, and c is a constant of the ligand to adjust the pK values so that all the values lie on the same scale. For hard acids, the value of Y is found to he > 2.8 and for soft acid > 3.2. For border line acids, the value ranges between 2.8 and 3.2. For example, the values of Y are as follows:

Li^+	Al^{3+}	Mg^{2+}	Na^+	K^+	Ca^{2+}	Fe^{3+}	Co^{3+}	Cs^+	Co^{2+}
0.36	0.70	0.80	0.93	0.93	1.62	2.32	2.56	2.73	2.96

Hard acids

Sn^{2+}	Tl^{3+}	Cu^{2+}	Pb^{2+}	Tl^+	Hg^{2+}	Au^+
3.17	3.2	3.45	3.58	3.78	4.25	5.95

Soft acids

The value of b (ligand parameter) also increases as we move from a hard base to a soft base.

OH^-	NH_3	Cl^-	Br^-	F^-	$S_2O_3^{2-}$
0.40	1.08	2.49	5.5	7.17	12.4

It is interesting to note that a and X are also regarded as another measure of hardness. However, these include the inherent acid base strengths also X has been found to he closely related to the electronegativity of the metal ion and is considered to be measure of the tendency of the metals ions to accept electrons from the ligands. Further X follows the same order as that William's order for the stability constants of the complexes

$$Mn(II) < Fe(II) < Co(II) < Ni(II) < Cu(II)$$

2. **π Bonding theory :** J. Chatt's pi-bonding theory can be applied to explain the bonding between the soft-soft complexes. According to this theory, the soft acids have loosely held outer d-orbitals electrons which can form pi-bonds by donating to suitable ligands. Such ligands are those in which empty d-orbitals are available on the basic atoms, eg. P, As. S, I, etc. The presence of d-orbital, on the ligands (except on CO) helps to strengthen the n-bonding.

3. **Pitzer's theory** : This theory of London and van der Waal's forces explains that the forces between the interacting soft-soft groups depend upon the product of polarizabilities of the interacting groups. Both soft acids and soft base have larges values of polarizabilities.

Electronegativity and Hardness and Softness : Hardness and softness have been found to be related to electronegativity. In brief, highly electronegative species are hard acids while lower electrongative species are soft acids. For example, highly electronegative ions like Li^+, Na^+, etc. are hard acids while transition metal ions having low electronegativities like Cu^+, Ag^+, Cu^+, etc. are soft acids.

Acid and Base Strength of SHAB : Any acid or base may be considered as hard or soft by its apparent preference for the hard and soft reactants. For instance, a base B is considered to be hard or soft depending upon the value of equilibrium constant K for the following reaction:

$$BH^+ + CH_3Hg^+ \rightleftarrows Ch_3HgB^+ + H^+ \qquad ...(1)$$

$$BH^+ \rightarrow B + H^+$$

$$k = \frac{[CH_3HgB^+][H^+]}{[CH_3Hg^+][BH^+]} = \frac{[CH_3HgB^+]}{[CH_3][Hg^+][B]} \frac{[H^+][B]}{[BH^+]} \qquad ...(2)$$

$$= K_s K_a \qquad ...(3)$$

or

$$\log K = \log K_4 + \log K_a \qquad ...(4)$$

If the value of K is less than unity, B is termed as a hard base and if it is greater than unity, B is termed as a soft base.

In equation (4), K, denotes the formation constant of CH_3HgB^+ from CH_3Hg^+ and B while K_a denotes the acid dissociation constant of the conjugate acid BH^+. From the values of K and K_a, it is also possible to decide whether B is hard or soft. If pK_a is more than log K_s, B is termed as hard base and if pK_a is less than log K_s, B is termed as soft base.

From the equation (2), it also follows that the reaction (1) will go to the left if B is a hard base. However, it will go to the right, if B is a soft base.

Now the question arises : Why methyl mercury (II) cation is selected in the above discussion 7 The simple reasons for this are :

(i) It is a typical soft acid,

(ii) It is univalent like proton, making the calculations simpler.

Hardness and softness arc quite different from the inherent acid-base strengths because these simply refer to the hard-hard and soft-soft interactions. For example, both hydroxide ion and fluoride ion are hard bases. However, the hydroxide ion (whose pK_a is 15.7) is 10^{13} times more basic than the fluoride ion (whose pK_a, is 2.85).

Certain examples are known in which a strong acid or base may displace a weak acid or base although this may violate the principle of SHAB interactions. For example, a stronger soft base sulphite displaces a weaker hard base fluoride from a hard acid proton:

$$SO_3^{2-} + HF \rightarrow SO_3^- + F^-, K_{eq} = 10^4$$

Similarly, a very strong hard base displaces a soft weak base from a soft methyl mercury (II) cation.

$$OH^- + CH_3HgSO_3^- \rightarrow CH_3HgOH + SO_3^-, K_{eq} = 10$$

The above replacement reactions take place against the hardsoft principle. However, when both the strengths and hardness are taken into cansiderations, the hard-soft rule works well. For example, consider the following reactions:

$$CH_3HgF + HSO_3^- \rightarrow CH_3HgSO_3^- + HF\ (K_{sq} = 10^3)$$

$$CH_3HgOH + HSO_3^- \rightarrow CH_3HgSO_3 + H_2O\ (K_{sq} = 10^7)$$

In the above reactions both sulphite (pK_a = 14.2) and hydroxyl ion (pK_a = 15.7) combine with protons. But the pK_s for sulphite ions with methyl mercury (II) cation is 21.2 while in case of hydroxyl ion is only 9.4 Therefore, sulphite is considered to be a soft base.

Applications' of the SHAB principle : The principle of soft and hard acid-base finds applications in various domains of chemical reactions. Some of them are:

(i) The simplest reaction that can be explained on this concept is

$$LiI + CsF \rightarrow LiF + CsI$$

This is an exothermic reaction with 13.9 kJ/mole heat of reaction. Here soft iodide prefers to combine with the soft caesium ion. The hard fluoride ion combines with hard lithium ion.

(ii) For heterogeneous catalysis, SHAB principle says that soft metals adsorb soft bases. For solubility, the rule is that hard solutes dissolve in hard solvents and soft solutes dissolve in soft solvents.

(iii) *Occurrence of minerals* : The kinds of metal ores found in the earth's crust can be rationalized by using this principle Thus, hard acids such as Ca^{2+}, Mg^{2+} and Al^{3+} appear as $CaCO_3$, $MgCO_3$ and Al_2O_3 respectively. The anions CO_3^{2-} and O^{2-} are also hard. These three hard acid cations are never found as their sulphides since S^{2-} is a soft base. However, soft acid such as Cu^+, Ag^+ and Hg^{2+} are found to be combined with the soft base S^{2-} as sulphides. The intermediate acids such as Ni^{2+}, Pb^{2+}, and Cu^{2+} are found as both sulphides and carbonates.

(iv) In both BF_3 and BH_3, the boron is trivalent but quite different behaviour can be explained on the basis of SHAB principle. The presence of hard fluoride ions in BF_3 makes it easy to add other hard bases, eg. $H_3N \rightarrow BF^3$ the presence of soft hydride ions in BH_3 makes it easy to add other soft bases. In BF_3 the boron is nearly B^{3+}. But the soft hydride ions donate negative charges extensively to boron, so that the boron atom is almost neutral and becomes soft.

(v) For elements of variable valency there is usually an increase in hardness with increase in oxidation state. The nickel (0) in. Ni $(CO)_4$ is soft, mickel (II) is borderine and nickel (IV) is hard.

(vi) The sulphur end of the thiocyanate ion is assumed to be softer than the nitrogen end, and hence prefers soft Lewis acids.

(vii) *Relative stabilities of Compounds* : By using Tables I and II and Pearson's rule, it becomes possible to explain the relative stabilities of the compounds and complexes. For example, AgI_2^- is stable, but AgF_2^- does not exist. As described earlier, Ag^+ is a soft acid F^- is a hard base and I^- is a soft base. Therefore, AgI_2^- (soft acid + soft base) is a stable complex and can explain that CoF_6^{2-} (hard acid + hard base) is more stable than CoI_6^{2-} (hard acid + hard base).

By using the SHAB principle, it is possible to predict the relative strength of halogen acids in aqueous medium. For example.

$$HX + H_2O \rightarrow H_3O^+ + X$$

(where X = F , Cl , Br or I)

In the above reaction, the hard acid H+ combines with the hardest base F to form the dissociated or weakest acid. Thus, the strengths of the halogen acids will follow the following order:

$$HF < HCI < HBr < HI$$

Weakest acid   strongest acid

Out of the two complexes, $[Cd(NH_3)_4]^{2+}$ and $[Cd(CN)_4]^{2-}$, later should be more stable (combination of soft acid Cd^{2+} with soft base (CN^-) on the basis of SHAB principle.

(iii) *Poisoning of metal catal catalysts* : This has been explained on the basis of SHAB principle. Soft metals (e.g Pd and Pt) act as catalysts and they are easily poisoned with CO, olefins, phosphorus or arsenic ligands (all soft bases). These ligands adsorb strongly on the surface of the metal and block the active sites. The soft metal catalysts are not affected by hard bases such as F.O.N.

(ix) *Rates of chemical reactions* : The SHAB principle has been used to correlate the rates of chemical reactions. The rates of electrophilic and nucleophilic reactions have been found to depend upon the hardness or softness of the acid or base.

(x) *Course of reactions* : The SHAG principle has been used to predict the course of a reaction. Consider the following reactions:

$$H^+ + CH_3HgOH \rightarrow H_2O + CH_3Hg^+ \quad \text{...(1)}$$

$$H^+ + CH_3HgSH \rightarrow H_2S + CH_3Hg^+ \quad \text{...(2)}$$

According to SHAB principle, the reaction (1) should move towards right because the hard acid H^+ is combining with hard base OH^- the reaction (2) should move towards left because the soft methyl mercury CH_3Hg^+ acid is combining with soft HS^- base.

Limitations of SHAB Concept

The various limitations of this concept are as follows:

(1) It does not involve any quantitative scale of measurement.

(ii) In order to explain the observed phenomenon, one has to break a compound to acid and base fragments. This breaking will be correct only if the reaction is known. For example, the break I for the scarification reactions explains the observed behaviour. However, the break 11 cannot be ignored and should be taken into consideration.

$$CH_3CO^+OH^- \rightarrow \quad C_2H_5O^-H^+ \text{ Break I}$$

$$CH_3COO^-H^+ \rightarrow \quad C_2H_5^+OH^- \text{ Break II}$$

(iii) In this concept, it is assumed that hard-soft factors do not depend upon the acidic or basic character of the compounds and therefore they work independently. However, many examples are known which show interdependence of the two concepts.

(iv) Due to the soft-soft (CH_3^+–H^-) interactions, it is expected that the following reaction should proceed.

$$CH_3^+ (g) + H_2(g) \rightarrow \quad CH_4 + H^+(g)$$

However, the unfavourable entropy change (3600 kJ mole^{-1}) indicates that this reaction should not proceed at all and the reason for this is that there is greater acidity of the proton relative to the CH_3 nucleus.

Conclusion

The hard-soft classification is a useful concept, but it does not lead directly to a scale of acid-base strength. Inherent acid-base strengths are also not taken into consideration. Thus the OH^- and F^- ions are both hard bases. But the former is nearly 10^{13} times as strong a base as the latter. Various approaches to correlate inherent acid-base strength with attributed hardness (and softness) factors are yet to be widely accepted.

Interpretation of many reactions by splitting the participants into acid-base fragments is also arbitrary to some extent. The very common reaction between ethanol and acetic acid may be interpreted in either of two ways:

$$CH_3COO^-H+ + C_2H_5 + OH^- \text{ or } CH_3CO + OH^- + C_2H_5O^-H^+$$

Either interpretation is justifiable by the hard-hard combination of H^+ with OH^-. The actual involvement of the second mechanism has to be understood by other means.

Sometimes the hard-soft principle fails to keep parity with inherent acid-base strengths. The reaction:

$$CH_3^+ (g) + H2(g) \rightarrow CH_4(g) + H^+(g)$$

should be favourable in view of the soft-soft combination of CH_3^+ with H^-. But actually the combination in endothermic by about 360 kJ per mole. (This has to be explained by the greater acidity of H^+ relative to CH_{3+}).

Similarly, hard-soft combinations take place in many cases like

$$SO_3^{2-} + HF \rightarrow HSO_3 + F^-$$

Here the soft bane SO_3^{2-} replaces the hard base F^- and combines with the hard acid H^+. The soft SO_3^{2-} is thus stronger than the hard F^-

However, we should restrict our expectations in view of the statement made by Pearson. "It should be stressed that the SHAB principle is not of theory but is a statement about experimental facts. Accordingly an explanation of some observation in terms of hard and soft behaviour does not invalidate some other, theoretical explanation."

5

Analysis of Chelates Compounds

INTRODUCTION

Chelate compounds are defined as those complexes in which the donor atoms are attached to each other as well as to the metal so that the metal becomes part of a heterocyclic ring.

Simple examples of complex formation and chelation may be represented schematically in the following way.

$$\begin{array}{cl} & \\ \text{M} + 4\text{Ä} \rightarrow \begin{array}{c} \text{A} \\ \downarrow \\ \text{A} \rightarrow \text{M} \leftarrow \text{A} \\ \downarrow \\ \text{A} \end{array} & \quad \text{Complex} \end{array}$$

$$\begin{array}{cl} \text{M} + 2\text{Ä} \rightarrow \begin{array}{c} \text{A} \\ | \\ \text{Ä} \rightarrow \text{M} \rightarrow \text{A} \\ | \\ \text{A} \end{array} & \quad \text{Metal chelate} \end{array}$$

where M represents the metal ion, A represents a complexing reagent and AùA represents a chelating reagent.

LIMITATIONS OF CHELATES DEFINITION

(i) According to the above definition, there is no restriction on either the size of the ring or the number of such rings.

(ii) The formal distinction made between a metal complex and a metal chelate compound seems rather arbitrary, and on this basis it is difficult to see how a metal chelate can differ very much from the metal complex containing the same kind of donor group.

CHARACTERISTICS OF CHELATES

1. They do not exhibit any of the properties of the salts. It is because that central metal atom has valency more than one. Chelates possess characteristics different from salts which are:
   (a) They have low melting points.
   (b) They are quite stable.
   (c) They diSSolv easily in organic solvents.
2. The donor groups are linked firmly to the metal ions present in the ring. However, it is covalent or coordinated covalent linkage which holds the metal ion into the chelating ring.
3. The presence of heterocyclic ring is essential for the formation of chelates.
4. The metal ion must have a suitable size.
5. The charge on the metal ions is generally large which is responsible for attracting the lone pairs of the electrons of the donor atoms.
6. In most of the chelates, the number of atoms forming the ring of the chelate is five or six but the former is more stable.

ORIGIN OF THE WORD CHELATE

The term 'chelate' was taken from the Greek word 'Chela' meaning crab's claw. This analogy is obvious because both the claw and ligand may hold an object (*i.e.*, the metal ion) through more than one point of attachment.

DISTINCTION BETWEEN METAL CHELATES AND METAL COMPLEXES

In a metal complex, the metal combines usually with an electron donor, not giving rise to cyclic rings.

$$\begin{array}{c} \mathrm{x} \\ \downarrow \\ \mathrm{M} + 4\ddot{\mathrm{x}} \rightarrow \mathrm{x} \rightarrow \mathrm{M} \leftarrow \mathrm{x}\ (\text{Metal complex}) \\ \downarrow \\ \mathrm{x} \end{array}$$

where $\ddot{x}$ is a **complexing agent**. On the other hand in metal chelate the metal combines with organic or inorganic molecules or ions which may use two-atoms donor, giving rise to cyclic rings.

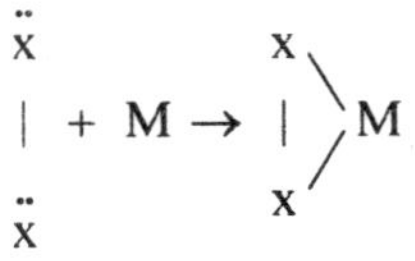

$$\begin{array}{c}\ddot{x}\\ |\\ \ddot{x}\end{array} + M \rightarrow \begin{array}{c}x\\ |\\ x\end{array}\!\!\!>\!M \text{ (Metal Chelate)}$$

where $\ddot{x}—\ddot{x}$ is **chelating agent**.

CHELATE RING FORMING GROUPS

The rings may be closed by the formation of covalent linkages, or coordinate linkages or by a combination of both.

(a) **Covalent Linkage :** Covalent linkage is produced by the replacement of a proton in an organic group. Functional groups of this type are sometimes called acidic groups because of the fact that hydrogen may be replaced from them. The most common examples of these are:

(i) — COOH (carboxyl group)

(ii) — SO_3H (sulphonic group)

(iii) — OH (enolic hydroxyl group)

(iv) = NOH (oxime).

(b) **Coordinate Linkage :** Coordinate linkages, without the replacement of hydrogen, are formed by the donation of an electron pair. The most common functional groups which contain donor atoms are:

(i) — NH_2 > NH and ≥ N (primary, secondary and tertiary amines)

(ii) = NOH (oxime)

(iii) — OH (alcoholic hydroxyl)

(iv) > C = 0 (carbonyl)

(v) —S— (Thioether).

CLASSIFICATION OF CHELATES

Dielh has classified chelates on the basis of:

(i) The total number of donor groups in the complexing agent *i.e.,* monodentate, bidentate, etc., and

(ii) The relative number of acidic and coordinating groups involved in ring formation.

Compounds which can link themselves with the central metal atom or ion through two bonds are called *two-toothed bidentate;* those capable of forming three bonds are known as *three-toothed* or *tridentate;* those forming four bonds are *quadridentate* and so on. This classification takes into consideration both the number of bonds and types of bonds as given below:

1. **Bidentate groups :** Examples are oxalate, glycine and phenanthroline.

(a) **Two acidic groups :** Rings of this type are formed by the replacement of two hydrogen atoms by the metallic cation, *e.g.*, platinum ions react with oxalate ions to form platinum oxalate.

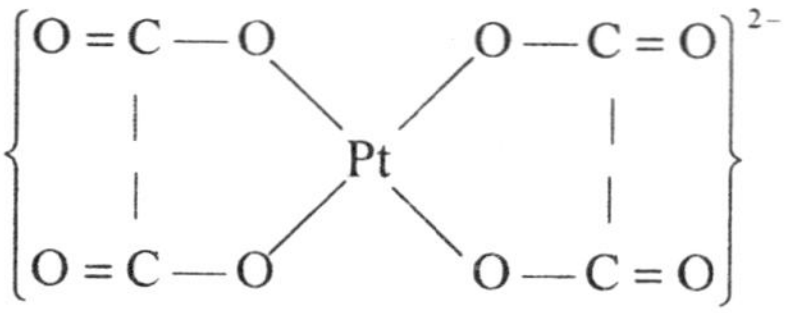

$$\left[\begin{array}{ccccc} O=C-O & & & & O-C=O \\ | & \diagdown & & \diagup & | \\ & & Pt & & \\ | & \diagup & & \diagdown & | \\ O=C-O & & & & O-C=O \end{array}\right]^{2-}$$

(b) **One acidic group and one coordinating group :** The cupric ion reacts with glycine to produce copper glycinate in which each glycine molecule is attached to the copper atom by one covalent bond and one coordinate bond.

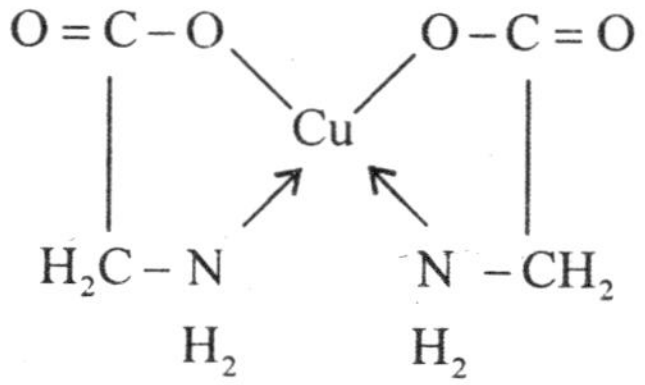

$$\begin{array}{ccccc} O=C-O & & & & O-C=O \\ | & \diagdown & & \diagup & | \\ | & & Cu & & | \\ | & \nearrow & & \nwarrow & | \\ H_2C-N & & & & N-CH_2 \\ H_2 & & & & H_2 \end{array}$$

(c) **Two-co-ordinating groups :** Orthophenathroline is a heterocyclic organic compound having two ring of nitrogen atoms. This substance forms a complex ion with the ferrous ion in which three molecules of the orthophenanthroline are attached to one ferrous atom by two co-ordinating linkages. The formula of this chelate is given below :

Tridentate Groups

(a) **Three acidic groups :** In tartar emetic, antimony atom is attached to the tartrate ion by the removal of three hydrogen atoms to form three covalent linkages.

$$\left\{\begin{array}{l} O=C-O \\ \quad | \\ H-C-O-Sb \leftarrow OH_2 \\ \quad | \\ O_2C-CH-O \end{array}\right\}^{-}$$

(b) **Two acidic groups and one co-ordinating group :** Aspartic acid combines with trivalent cobalt to form a water soluble tridentate compound. The structure given to this complex ion is

$$\left\{\left(\begin{array}{l} \quad CO---O \rightarrow \\ CH_2 \\ \quad CH-H_2N \rightarrow Co \\ \quad | \\ \quad CH---O \rightarrow \end{array}\right)_2\right\}^{-}$$

(c) **One acidic group and two co-ordinating groups :** The cobalt ion interacts with diaminopropionic acid to give a tridentate compound in which each molecule of the acid is attached to the cobalt atom by the two co-ordinating linkages and one covalent linkage. The formula for this chelate is

$$\left(\left[\begin{array}{l} O=C--O \rightarrow \\ \quad | \\ H-C-NH_2 \rightarrow Co \\ \quad | \\ H-C-NH_2 \rightarrow \\ \quad H \end{array}\right]\right)^{+}$$

(d) **Three co-ordinating groups :** Triaminopropane combines with cobalt (III) to produce a chelate, in which one molecule of the triaminopropane occupies three co-ordination positions about the metallic atom. The structure for this is

```
⎧⎛        H         ⎞     ⎫ 3+
⎪⎜  H − C − NH₂ ↘    ⎟     ⎪
⎪⎜      |             ⎟     ⎪
⎨⎜  H − C − NH₂ ──→ Co ⎟   ⎬
⎪⎜      |          ↗  ⎟     ⎪
⎪⎜  H − C − NH₂      ⎟     ⎪
⎩⎝        H         ⎠₂    ⎭
```

4. **Quadridentate and sexadentate groups :** Very few polydentate chelating agents beyond those capable of filling tridentate groupings are known

   (a) **Quadridentate Groups :** Two of the possible five classes are known:

      (i) ***Two acidic groups and two co-ordinating groups :*** Ethylene diamine-acetic acid reacts with the cupric ion to produce a quadridentate grouping containing two covalent bonds and two coordinate linkages. The structure of this compound is:

```
        CH₂───CH₂
       /         \
     NH           NH
    /   \       /   \
 H₂C     ↘   ↙     CH₂
  |                  |
 H₂C       Cu        CO
    \     /   \     /
      O          O
```

      (ii) ***Four co-ordinating groups :*** Tetrapyridyl reacts with certain bivalent transition metals, such as Fe^{2+}, Co^{2+}, Ni^{2+}, Cu^{2+} and Zn^{2+} to form a quadridentate compound in which the organic molecule is attached to the metal by four co-ordinating linkages. In the case of Fe^{2+}, the structure of the chelate is given.

   (b) **Sexadenate groups :** In 1950 Dwyer and Lions have reported several organic molecules containing six groups capable of attaching themselves in six octahedral positions of the co-ordinating sphere of a central metallic atom. One of the molecules is 1, 8-bis-salicylideneamino-3, 6-dithiooctane which reacts with zinc, nickel, copper or cobalt to form a chelate compound with the possible structure.

FACTORS INFLUENCING THE STABILITY OF METAL CHELATES

1. **Size of the chelate ring :** In general it has been reported by many workers that saturated compounds tend to form five membered structures while those ligands which contain double bonds tend to form six-membered rings. Ley from conductometric studies of amino acid chelates concluded that chelates, with five and six membered rings are most stable. According to Schawarzenbach maximum chelate stabilisation is attained when a strainless rigid ring of a size, that can just accommodate the metal ion, is formed.

2. **Number of rings within the chelate :** It has been known for long that increase in the number of rings within the chelate results in greater stability. Quantitative studies have confirmed this. It is also found that the formation of fused rings around the metal tends to provide greater stability than the formation of single rings. For example, the chelate (I) having three interlocked ring system (fused rings) is much more stable than the chelate (II) involving comparable two ring system.

3. **Resonance effect :** The importance of resonance effects on the stability of metal chelates has been investigated by Calvin and Wilson (1945). They came to the conclusion that unhindered resonance of the chelate ring enhances the stability. On the other hand resonance in the attached benzene rings may interfere with resonance of the chelate ring and in such cases, stability of the chelate is lowered down.

4. **Steric effect :** This effect modifies the stability of the metal chelate to a great extent. Morgan found that acetyl-acetone did not give the usual coloured chelates with cupric or ferric ion when the 3-position was substituted by an isopropyl or secondary butyl group, but gave typical chelates when straight chain alkyl groups were present. This seemingly strange behaviour is due to steric effect of isopropyl and secondary butyl groups.

5. **Basic strength and chelate stability :** The more available the electron pair of the donor group the stronger is the covalent bond which it may form with a metal.

6. **Effect of substitution on the ligand :** The substitution of a group on the chelating agent may influence the tendency for chelation in one or two ways:

   (a) It may influence the acidity of the donor groups or may interfere with or enhance the resonance of the chelate rings

   or

   (b) The addition of groups on the ligand due to purely steric effects may prevent the ligand ions of molecules from acquiring the orientation about the central metal ion. (most favourable for chelation). Substitution of an ethoxyl groups in place of a methyl groups of acetyl acetone greatly decreases the stability constant. This is due to the interference of the strong resonance of ethyl acetoacetate which prevents the carbonyl group from taking part fully in the resonance of the chelate ring.

7. **Effect of metal ion :** Transition metals usually form covalent bonds with electron donors. On the other hand, in the chelate compounds of alkaline earths and alkali metals, ions must form ionic bonds.

However, the stability of transition metal cheletes decreases with increasing basicity of the metal. Those transition metals which form the strongest covalent bonds with the ligand molecules produce stablest chelates.

The stability decreases with increasing basicity of the metal : Also metals which form the strongest homopolar bonds form the most stable chelates. For chelates of the primary ionic type such as those of the alkaline earth and alkali metals the strength of chelation should increase with increasing ionic charge for both metal ion and donor and increase with decreasing ionic radius of the metal ion.

DETECTION OF CHELATES

Most of the methods merely demonstrate complex formation. Thus, the, chelate nature of the complex ion is often infered by others, more or less indirect methods. Some of these methods are described below:

(a) **Absorption Spectra :** Many workers recognised that a characteristic colour is one of the most important and distinguishing features of chelate compounds, and used absorption spectra to distinguish between free metal ions and their chelates. Intensification of colour is usually considered an indication that chelate may be present.

Simple colorimetric methods are used to determine the compositions of chelate compounds in solution. If the chelate is the only coloured substance present, the optical density of the solution is proportional to chelate concentration. Hence a plot of optical density versus composition of the solution would give a curve with a maximum at the composition corresponding to the formula of the chelate.

(b) **pH Effect :** All metal chelates may be considered as formed by the displacement of one or more usually weak acidic protons of the chelating agent by the metal ion. Thus, the addition of glycine to a solution of a copper salt causes a drop in the pH. This is due to the following equilibrium.

$$Cu^{2+} + 2H_3\overset{+}{N}.CH_2COO^- \rightleftharpoons \begin{array}{ccccc} & & & H_2 & \\ CO-O & & & N--CH_2 & \\ | & \searrow & & \swarrow & | \\ | & & Cu & & | \\ | & \nearrow & & \nwarrow & | \\ CH_2-N & & & O--CO & \\ H_2 & & & & \end{array} + 2H^+$$

Thus, the chelate formation of metal ions with amino acids and with the acid form of other chelating agents, results in pH drop and the greater the tendency for metals to combine with a given chelating agent, the greater the drop in pH.

The pH effect can be used as an indication of the extent of chelation.

(c) **Optical Activity** - The resolution of chelate compounds into optically active forms has provided powerful evidence in the configuration of ring structures. For example, 1, 2, 3-triaminopropane can react with a metal so as to occupy only two coordination positions, the third amine group then being capable of salt formation. The complex compound of this type formed with $PtCl_4$ will then he either dissymmetric (a) or symmetrical (b), according as a 5- or 6-membered ring is formed preferentially through chelation. For example, Mann (1926) was able to resolve this complex chelate, thereby confirming the existence of the 5-membered ring (a).

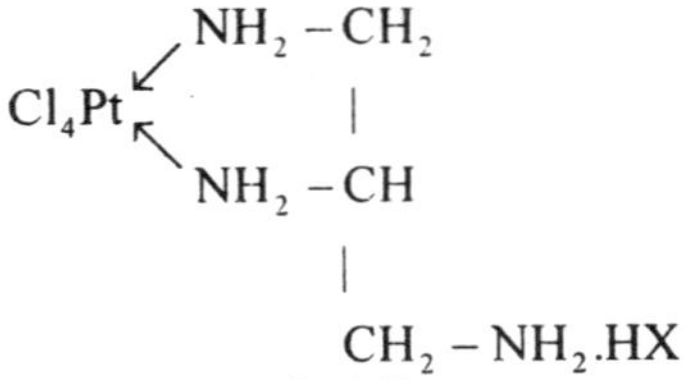

$$Cl_4Pt\begin{matrix}\swarrow NH_2-CH_2\\ \qquad\quad | \\ \nwarrow NH_2-CH\end{matrix}$$
$$\qquad |$$
$$CH_2-NH_2.HX$$

Resolvable
5-membered ring

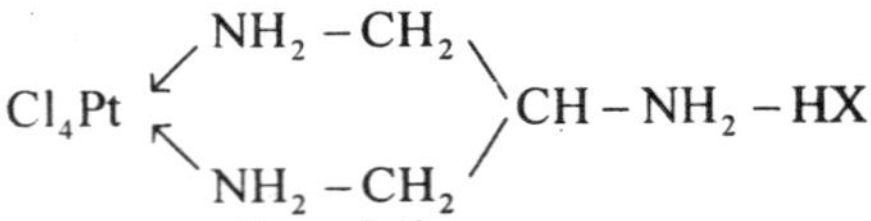

$$Cl_4Pt\begin{matrix}\swarrow NH_2-CH_2\diagdown \\ \nwarrow NH_2-CH_2\diagup\end{matrix}CH-NH_2-HX$$

Non-resolvable
6-membered ring

This method cannot be applied to chelate rings in general since not all chelate structures are optically active.

(d) **Conductance Measurements** - Conductance measurements constitute probably the earliest method to study the formation of chelates. The disappearing of ions in complex formation may he followed by electric conductance measurements of the solution in which they are formed.

Chelate formation does not always result in a decrease in electrical conductance, since it is possible to increase the conductance of the medium by a suitable reaction in which highly conducting ions are produced in the course of reaction, *i.e.,* chelate formation. For example, the addition of two molecules of ghycine to one molecule of aqueous copper acetate solution decreases the conductivity due to the formation of an uncharged chelated as

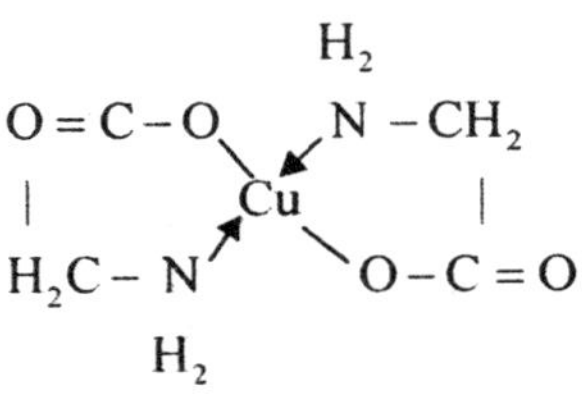

$$\begin{matrix} & & H_2 & \\ O=C-O & & N-CH_2 \\ | & Cu & | \\ H_2C-N & & O-C=O \\ H_2 & & & \end{matrix}$$

Acetic acid produced in this reaction is a weak electrolyte, contributing little towards electrical conductivity.

Thus, this method is not reliable and gives only indication about complex formation and does not provide a demonstration of chelation.

(e) **Polarographic Measurements** - Valuable information about the composition of metal chelates and their behaviour in aqueous solution has been obtained from a study of their reduction at the dropping mercury electrode.

The polarographic technique is based on the fact that the reduction potentials of metal ions are shifted to more negative values, as a result of chelate formation, and the correlation of this displacement of potential with both the concentration of the ligand and with solution, and pH provides an excellent method for determining the composition of metal chelate compounds.

(f) **X-ray Measurements** - The only direct proof for the existence of chelate ring is provided by X-ray crystal analysis. All other methods are indirect. The main difficulty lies in obtaining crystals possessing sufficient symmetry.

IMPORTANCE OF CHELATES

Although, the first chelate was recognized and described by Werner, the development and main importance of chelates has taken place almost entirely within recent years.

1. Among the earlier investigations Willstatter and H. Fischer, established the general structures of chlorophyll and hemin. Chlorophyll is a magnesium and hemin is an iron chelate.

2. **Analytical Chemistry**

   (a) **Qualitative Analysis :** Nickel, magnesium and aluminium in qualitative analysis are identified by the formation of stable, highly coloured chelates.

   Thus, dimethyl glyoxime, 8-hydroxy quinoline, o-phenanthroline, Fehling's solution and aluminon are used for the detection of many metal ions. For example, dimethylglyoxime reacts with Ni (II) to form an insoluble scarlet red precipitate. 8-Hydroxyquinoline is used for the determination and detection of over 30 elements such as Mg, Al, Zn, etc. The precipitates formed by this reagent with all these metals involve highly stable and characteristically coloured chelates.

(b) **Gravimetric Analysis :** Chelating agents are used in the gravimetric analysis of magnesium, nickel, etc., by their quantitative precipitation. The advantage of weighing the chelate is the high molecular weight of the precipitate and the low metal contents. This makes possible the determination of small amounts of metal with great accuracy.

(c) **Volumetric Analysis :** Chelating agents are frequently used as indicators in the titration of certain metal ions.

For example, ferroin, which is a chelate complex and contains $[Fe(phen)_3]^{2+}$ ion, is deep red in colour and acts as an indicator in oxidation-reduction titrations. In these titrations, the removal of the colour of the ferroin indicates that the end point has been reached. The ferroin is oxidised only after the reducing agent being titrated is completely oxidised.

3. **Water Softening :** Ion exchange resins have been produced which can successfully remove both cations and anions from aqueous solutions and the purity of water is nearly that of distilled water.

4. **Removal of Metal Ions** - The sequestering agents, which form water-soluble chelates, are used in aqueous solution for the effective removal of objectionable metal ions. These agents are also used in the removal of boiler scale during laundering.

5. **Elimination of Radioactive Metals :** Chelating agents such as ethylenediamine tetraacetic acid are used to speed up the elimination of harmful radioactive metals from the body. In using such agents, side effects, if any, should carefully be compensated.

For example, EDTA can be used to counteract the effect of lead poisoning but at the same time it may result in decalcification of bones and teeth because it (EDTA) forms a very stable calcium chelate. In order to prevent such decalcification, EDTA is used in the form of $Na_2CaEDTA$.

6. **For Solvent Extraction :** Chelates which form soluble metal chelates with precipitates are used for solvent extraction.

   For example, the separation of Zr and Hf was a difficult analytic task until it was found that the chelates of these two metals with theonyl trifluoroacetone possess different solubilities in benzene. The distribution coefficient of these two metals chelates differ by a factor of 20, *i.e.,* zirconium has a greater tendency to be extracted. Therefore, it is now possible to effect a good separation of these two metals at present.

7. **Food Preservation :** Fruits, fruit juices and food stuffs are preserved with the help of chelating compounds. The addition of one hundredth of one percent of sequestering agents such as EDTA improves the qualities of the food on standing by catching the traces of metals (which are responsible for spoiling the food) by chelate formation.

INTRODUCTION TO REDOX POTENTIAL

Electrode Potential

When a metal is kept in contact with a solution of its own ions, the metal exhibits a certain tendency to accept or lose electrons from solution. This tendency of a metal sets up an electrical potential difference between the metal and its solution. This is called electrode potential and is symbolised as E.

Standard Electrode Potential : When the concentration or activity of solution is unity, the difference of potential between the metal and the solution of its cations is called the standard electrode potential. This is symbolised as E°.

For ionic solution, the reference concentration is 1M. For gases, a pressure of 1 atm is chosen as the standard pressure.

Oxidation Potential and Reduction Potential : Each electrode is making a contribution to the cell potential which is called the electrode potential. Generally, the electrode potential has been regarded as the measure of the tendency of an electrode to gain or lose electrons.

The tendency to lose electrons is also meant the tendency to get oxidised. Therefore, this tendency is termed as oxidation potential. Similarly, the tendency to gain electron is the tendency to get reduced. Hence, this tendency is termed as reduction potential.

In order to understand the meaning of oxidation potential and redox potential, a strip of zinc is kept in an aqueous solution of zinc ions (produced by a zinc salt). This metal has a tendency to liberate electrons and forms Zn^{2+} ions in solution while Zn^{2+} ions already present in solution have a tendency to accept electrons and to be discharged as the metal.

$$Zn(s) \rightleftarrows Zn^{2+}(aq) + 2e^-$$

This is a reversible reaction in which oxidation of zinc metal into zinc ions occurs towards the right while reduction of zinc ions to zinc metal occurs towards the left. The equilibrium between the two opposite reactions gets established rapidly.

As zinc possesses a greater tendency to go into solution as zinc ions, it might have liberated some electrons before equilibrium is attained. These electrons get accumulated on the metal strip, thus building up a negative potential with respect to the solution. But this potential has been developed due to the tendency of the metal to get oxidised to its ions. Therefore, the potential is called the oxidation potential of the metal. Here the metal acts as a reducing agent.

If the reaction is taking place towards the left, Zn^{2+} ions will have the tendency to gain electrons and get converted into zinc metal. But the tendency to gain electrons implies the tendency to get reduced.

Therefore, this tendency is known as reduction potential of the Zn^{2+} ions. Thus, the potential developed on a metal when it is immersed in its ions is also known as oxidation-reduction potential or redox potential. Greater the tendency for a metal to liberate its electrons when it is kept in a solution of its own ions the greater will be its redox potential. Also, greater will be its tendency to act as, a reducing agent.

The magnitude of the redox potential has been found to depend upon the following facts:

(i) Tendency of the metal to ionise,

(ii) Concentration of the ions already present in solution and

(iii) Temperature.

NERNST'S EQUATION

This equation describes the quantitative relationship between electrode potentials and concentrations of the involved substances in the electrode or half-cell reaction. This relation is essential because galvanic cells are operating under concentration conditions different from the standard states conditions.

For reduction electrode potentials, the half-cell reaction should be written as the reduction reaction, *i.e.*,

$$A^{n+} + ne \rightarrow A$$

For the above reaction, Nernst deduced the following mathematical equation:

$$E = E^0 - \frac{2.303RT}{nF} \log \frac{[A]}{[A^{n+}]} \qquad \ldots(1)$$

Where E = Electrode potential of a given system $A^{n+} + ne \rightarrow A$,

E^0 = Standard electrode potential of given system,

R = Gas constant ($8.314 JK^{-1} mol^{-1}$),

T = Kelvin temperature (298 K, the temperature at which values are given),

F = Faraday constant ($96500\ C\ mol^{-1}$), and

n = Number of electrons involved (gained or lost) in the reduction reaction.

On substituting the values of R. T and F in the equation (1), we get

$$E = E^0 - \frac{0.059}{n} \log \frac{[A]}{[A^{n+}]}$$

Determination of Electrode Potential

It is not possible to measure single electrode potential because during its measurement whenever a second metal is put into an electrolyte solution, it becomes an electrode itself and we get a new galvanic cell. This difficulty in measuring single electrode potentials has been avoided by selecting one of the electrodes as a reference electrode and giving to it a potential of zero volt. Generally, the normal hydrogen electrode (NHE) has been universally accepted as a reference electrode and has been assigned the zero volt value.

NORMAL HYDROGEN GAS ELECTRODE

This consists of a glass jacket with a small inlet at the top and a number of outlets at the bottom.

A piece of platinum metal fused into glass tube is fitted inside the Jacket. Invariably the surface of the metal is coated with platinum black.

The glass jacket is immersed in the acid solution with a hydrogen ion concentration of one mole per litre (1M or 1N) and dry hydrogen gas at 1 atm pressure is bubbled through the solution. The gas is adsorbed on the platinum
surface. An equilibrium between the gas and hydrogen ions (H^+) is established across the metal.

$$\frac{1}{2}H_2(g) \rightleftarrows H^+(Sol) + e.$$

The electron set free is picked up by the platinum metal and transferred to the other electrode.

Measurement of Standard Reduction Potential

A simple scheme for determining standard reduction electrode potential using normal hydrogen electrode. In this figure, hydrogen electrode is shown on the left. Potentials of all others electrodes can he found by coupling them with NHE. The E.M.F. of cell is measured by the potentiometer or voltmeter and the measured E.M.F. will he equal to the required potential of the electrode on hydrogen scale by taking potential of hydrogen electrode as zero.

In every electrode, if the electrolyte is 1 molar in concentration, the solid phase consists of pure substance or substances, and the gas, whenever is used, is at a partial pressure of one atmosphere, this electrode is known as a standard electrode and its measured potential is known as the standard electrode potential. If standard electrodes are used, the E.M.F. is given the symbol E^0.

When the circuit is closed, the voltmeter gives a reading of 0.76 V. The pointer is deflecting towards the right. This reveals that conventional current flows from the hydrogen electrode towards the zinc electrode, *i.e.*, the electrons flow from the zinc towards the hydrogen electrode. The two half-reactions will be as follows :

Oxidation half-reaction : $Zn\ (s) \rightleftarrows Zn^{2+}\ (aq) + 2e^-$

Reduction half-reaction : $2H^+\ (aq) + 2e^- \rightleftarrows H_2\ (g)$

The complete cell reaction would be obtained by adding the two half-reactions:

$$Zn\ (s) + 2H^+\ (aq) \rightleftarrows H_2(g) + Zn^{2+}\ (aq)$$

Suppose a copper strip dipping in a solution of Cu^{2+} of concentration 1M is used in place of the zinc electrode dipping in a solution of Zn^{2+} of concentration 1M. In that case, the voltmeter reads 0.34 but the pointer deflects towards the left which indicates that the conventional current flows from copper to hydrogen electrode. Thus the electrons flow from copper electrode to hydrogen electrode. The two half-reactions will he as follows:

Oxidation half reaction: $H_2(g) \rightleftarrows 2H^+\ (aq) + 2e^-$

Reduction half-reaction : $Cu^{2+}\ (aq) + 2e^- \rightleftarrows Cu(s)$

and the complete cell reaction will be

$$H_2(g) + Cu^{2+}\ (aq) \rightleftarrows 2H^+ (aq) + Cu(s)$$

Hydrogen gas is converted into H^+ ions while Cu^{2+} ions get discharged as copper metal.

Sign of Electrode Potential

Suppose an electrode is combined with NHE. If it undergoes reduction process spontaneously, it is said to have a positive value of reduction electrode potential. In other words, it is said to have an equal but negative value of oxidation electrode potential.

In this, the negative sign is an indication that the oxidation at the electrode when coupled with NHE will not he spontaneous. By following the same argument, an electrode, which undergoes oxidation change when coupled with NHE, will have a positive value of oxidation electrode potential and a negative value of reduction electrode potential.

Let us now draw list of standard electrode potentials for the known electrode systems These values can be either standard oxidation electrode potentials or standard reduction electrode potentials.

The list prepared of these is of standard reduction electrode potentials. A list of common standard electrode potentials is given in Table 1.

As all the electrodes do not undergo spontaneous reduction change when coupled with NHE, the values given in Table 1, have positive or negative values.

Table 1 : Standard reduction electrode potentials at 298 K in water

Electrode	*E^0 Volts*	*Electrode Reaction*
Li; Li^+	–3.05	$Li^+ + 1e \rightarrow Li$
K; K^+	–2.95	$K^+ + 1e \rightarrow K$
Ba; Ba^{2+}	–2.90	$Ba^{2+} + 2e \rightarrow Ba$
Ca; Ca^{2+}	–2.87	$Ca^{2+} + 2e \rightarrow Ca$
Na; Na^+	–2.71	$Na^+ + e \rightarrow Na$
Mg; Mg^{2+}	–2.37	$Mg^{2+} + 2e \rightarrow Mg$
Al; Al^{3+}	–1.66	$Al^{3+} + 3e \rightarrow Al$
Zn; Zn^{2+}	–0.76	$Zn^{2+} + 2e \rightarrow Zn$
Fe; Fe^{2+}	–0.44	$Fe^{2+} + 2e \rightarrow Fe$
Pb, $PbSO_4$; SO_4^{2-}	–0.31	$PbSO_4 + 2e \rightarrow Pb + SO_4^{2-}$
Co; Co^{2+}	–0.28	$Co^{2+} + 2e \rightarrow Co$
Ni; Ni^{2+}	–0.25	$Ni^{2+} + 2e \rightarrow Ni$
Sn; Sn^{2+}	–0.14	$Sn^{2+} + 2e \rightarrow Sn$
Pb; Pb^{2+}	–0.13	$Pb^{2+} + 2e \rightarrow Pb$
Pt, H_2; H^+	±0.00	$2H^+ + 2e \rightarrow H_2$
Cu; Cu^{2+}	+0.34	$Cu^{2+} + 2e \rightarrow Cu$
Ag; Ag^+	+0.80	$Ag^+ + 1e \rightarrow Ag$
Cl^-; Cl^2, Pt	+1.36	$Cl_2 + 2e \rightarrow 2Cl^-$
Au; Au^{3+}	+1.50	$Au^{3+} + 3e \rightarrow Au$
F^-; F_2Pt	+2.87	$F_2 + 2e \rightarrow 2F^-$

ELECTROCHEMICAL SERIES

This is a series in which the electrodes are arranged in the decreasing potential value on the hydrogen scale. The electrode with the most negative values is at the bottom.

For metals, electrochemical series is also known as activity series. Sometimes, electrochemical series is also known as E.M.F. series.

Applications : The following generalised conclusions may be drawn from the electrochemical series:

1. To predict whether a metal will liberate hydrogen from a hydracid or not

All couples lying above the H_2/H^+ couple, *i.e.*, those with negative reduction potentials, act as reducing couples Hence, all metals (*e.g.*, zinc and iron) lying above hydrogen in the electrochemical series displace H_2 gas from a dilute acidic solution, *i.e.*, they themselves pass into solution as their ions and yield the appropriate salt. For example,

$$Mg + H_2SO_4 \rightarrow MgSO_4 + Hg$$

$$Mg \rightarrow Mg^{2+} + 2e$$

$$2H^+ + 2e \rightarrow H_2$$

$$Mg + 2H^+ \rightarrow Mg^{2+} + H_2$$

Conversely no metals (*e.g.*, copper and silver) lying below hydrogen in the electrochemical series are capable of reacting with dilute acids to form salts with the liberation of hydrogen.

2. To compare the relative tendencies of metal to change into cations

Any metal in the electrochemical series will reduce the ions of any metal lying below it while its ions are reduced by any metal lying above it. For example,

$$Zn + Cu^{2+} \rightarrow Zn^{2+} + Cu$$

$$Cu + 2Ag^+ \rightarrow Cu^{2+} + 2Ag$$

3. To compare the reactivities of non-metals

Any halogen in the electrochemical series will oxidise the ions of these halogens lying above it. For example,

$$Cl_2 + 2Br^- \rightarrow 2Cl^- + Br_2$$

$$Cl_2 + 2I^- \rightarrow 2Cl^- + I_2$$

$$Br_2 + 2I^- \rightarrow 2Br^- + I_2.$$

4. Calculation of the voltage of a galvanic cell

The E.M.F. of any cell can be computed from the values of standard electrode potentials taken from electrochemical series by employing the following relationship :

$$E_{cell} = E_{right} - E_{left} = E_{cathode} - E_{anode} \quad \ldots (1)$$

The above relation is only valid for the following two conditions:

(i) In the galvanic cell, anode is on the left hand side while cathode is on the right hand side.

(ii) The half-cell potentials of both E_{right} and E_{left} electrodes are reduction electrode potentials.

If the positive value of E_{left} is obtained, this means that the cell reaction is spontaneous.

If the negative value of E_{cell} is obtained, it means that cell reaction is non-spontaneous. However, if the cathode of the cell is made the anode and the anode is made the cathode, the reaction becomes spontaneous, *i.e.*, reverse cell will start working.

5. Rusting of iron

One can explain the phenomenon of rusting of iron (which is an electrochemical process) by employing electrochemical series. All the commercial forms of iron and steel are not pure but contain impurities, *i.e.*, these are heterogeneous.

But the different parts of the metal (iron or steel) possess different solution pressures. Therefore, the electrochemical couples are readily set up and there occurs the dissolving of metal due to local action.

6. Chemical behaviour

The electrochemical series offers a means of predicting the chemical behaviour of different elements. Thus, the alkali metals which occupy the top positions of the electrochemical series are the most reactive of all metals.

EMF DIAGRAM

It is possible to express the redox potentials of elements in different oxidation states graphically in abbreviated form in the EMF diagram. This is a short form of representing the redox reactions for an element in different oxidation states in which ions in the medium (H^+, OH^-), and molecules of the medium (H_2O) are ignored. Consider the following half-cell reactions

$$V^{2+} + 2e^- \rightleftarrows V \qquad E^0 = -1.186\ V$$

$$V^{3+} + e^- \rightleftarrows V^{2+} \qquad -0.256\ V$$

$VO^{2+} + 2H^+ + e^- \rightleftharpoons V^{3+} + H_2O$ +0.359 V

$V(OH)_4^+ + 2H^+ + e^- \rightleftharpoons VO^{2+} + 3H_2O$ +1.00 V

$V(OH)_4^+ + 4H^+ + 5e^- \rightleftharpoons V + 4H_2O$ -0.254 V

...(1)

which can be represented as the EMF diagram for vanadium as

$$E_A^0 \quad V(OH)_4^+ \overset{1.00}{-} VO^{2+} \overset{0.359}{-} V^{3+} \overset{-0.256}{-} V^{2+} \overset{-1.186}{-} V \quad (V(OH)_4^+ \text{ to } V: -0.245) \qquad ...(2)$$

In an EMF diagram, the species have been arranged from left to right in the order of decreasing oxidation numbers and the appropriate reduction potentials are inserted between the oxidized and reduced species as shown in Eq. (2). Similarly for manganese, the EMF diagram is

$$E_A^0 \quad MnO_4^- \overset{0.56}{-} MnO_4^{2-} \overset{2.26}{-} MnO_2 \overset{0.95}{-} Mn^{3+} \overset{1.51}{-} Mn^{2+} \overset{-1.18}{-} Mn \quad (MnO_4^- \text{ to } MnO_2: 1.70;\ MnO_2 \text{ to } Mn^{2+}: 1.23)$$

...(3)

As desired, it is possible to draw the EMF diagrams in the acidic, neutral or the basic medium as desired. The condition is denoted by putting the letters, E_A^0, E_N^0 or E_B^0 respectively before the diagram. Thus for manganese in neutral medium, the EMF diagram changes to.

$$E_N^0 \quad MnO_4^- \overset{0.56}{-} MnO_4^{2-} \overset{1.43}{-} MnO_2 \overset{-0.73}{-} Mn^{3+} \overset{1.51}{-} Mn^{2+} \overset{-1.18}{-} Mn \quad (MnO_4^- \text{ to } MnO_2: 1.33)$$

...(4)

It is possible to depict graphically the redox relationship between the different oxidation states of an element in yet another form. The free energy changes (in KJ mol^{-1} or in eV) for the reaction

$$M^{n+} + ne^- \rightleftharpoons M$$

(given as –nFE°) can be plotted against the oxidation state n readily. The lowest point on the graph denotes the most stable state. The negative ΔG values (or positive E^0 values) are indicated by the downward (negative) slope of the line joining the two oxidation states while positive ΔG values (negative E^0 values) are indicated by the upward slope of the line. A state on the convex point is unstable towards disproportionation (*e.g.*, Mn (VI). and Mn(III) for manganese).

Although this diagram is useful in projecting a general picture of the aqueous chemistry of the element in different oxidation states, such oxidation state diagrams lose their significance because

(i) The cell potentials E, and hence the free energy changes are pH dependent;

(ii) Complexation reactions may take place in solutions, especially in the presence of complexing ions (including halide ions),

(iii) Precipitations may take place completely altering the E values and the free energy changes. In such cases, a three-dimensional model is needed to demonstrate all these changes, thereby making the drawings very cumbersome.

PERIODIC TRENDS IN EMF VALUES

It is better to understand the variations of the E^0 values more clearly by considering the enthalpy change or the heat released during the oxidation of the metal to its ion (in form of the cycle).

This yields

$$Q = -\Delta H = -S-(I_1 + I_2 + \dots + I_n) + \Delta H_{hyd}.$$

where S refers to the sublimation energy of the solid metal, I the ionization energy, ΔH_{hyd} the hydration energy of the ion M^{n+} and Q the heat released during the process. If the entropy of the last step is regarded to be negligible, the emf (E°) of the oxidation process ($-\Delta G^0$/nF will be proportional to the heat envolved of (Q). As the hydration energy for most of the ions has been similar (except for Li^+ for which the hydration energy is very large), the trends in S and I, in general, ascertain the trends in the E° values. The E° values are seen to change as given below :

(1) Within a period, E^0 decreases because of the increase in the ionization energy with increasing atomic number.

(2) For an element, E^0 decreases with increasing oxidation number because of the increase in the $\sum I/n$ value with increasing n.

(3) In the transition metal series both S and I get increased with increasing atomic number, resulting in the decrease in the E^0 values in a period.

(4) Oxyanions, utilizing outer d orbitals (SO_4^{2-}, SeO_4^{2-}), become stronger oxidizing agents with increasing atomic number but those using their lower d orbitals show decrease in the oxidizing strength with increasing atomic weight.

EMF and the Electronegativity

If we couple the following reaction

$$HX \rightleftarrows \frac{1}{2}X_2 + H^+ + e^-$$

with the reaction

$$H^+ + e^- \rightleftarrows \frac{1}{2}H_2$$

yields

$$HX \rightleftarrows \frac{1}{2}H_2 + \frac{1}{2}X_2$$

for which, from Pauling's electronegativity X,

$$AH = \sim 106.4(X_x - X_H)^2 KJ\ mol^{-1}$$

if solvation effects are not taken into consideration.

Further, if the entropy changes are also ignored,

$$-E^0 \propto (X_x - X_H)^2 \qquad ...(2)$$

This relation can be utilized to rationalise the trends that:

(i) For the compounds the oxidizing strength tends to increase as follows:

$H_2N - NH_2 < HO - OH < F - F$

$H_2P - PH_2 < HS - SH < Cl - Cl$

$F_2 > Cl_2 > Br_2 > I_2$.

(ii) For the hydrides the reducing strength has been found to follow the following order:

$CH_4 < NH_3 < H_2O < HF$

$HF < HCl < HBr < HI$

$H_2O < H_2S < H_2Se < H_2Te$

$NH_3 < PH_3 < AsH_3 < SbH_3$.

Equation (2) is used for the evaluation of the group electronegativities from the emf data.

INTRODUCTION TO BORAZOLE

The most important ring system of organic chemistry is the benzene ring either as a separate entity or in polynuclear hydrocarbons such as naphthalene, anthracence, and phenanthrene. Inorganic chemistry has two (at least) analogues of benzene: borazine and trimeric phosphonitrilic compounds.

Borazine, B_3N_3He, is of special interest because it is isoelectronic with benzene and some of its physical properties show some striking similarities with this (Table 2). Indeed, the earlier name borazole was proposed for borazine because of the resemblance to benzene (benzol) and it has been frequently referred to as inorganic benzene. The systematic name is s-triazatriborine, the compound is still known as borazine.

Table 2 : Comparison of the physical properties of Borazine and Benzene

Compound	*Mol wt.*	*M P. °C*	*B P. °C*	*Δ Hvap KJ mole^{-1}*	*Trouton's constant*
Borazine	80.6	–58	64.5	29.3	21.4
Benzene	78	+6	80.0	31.0	21.4

METHODS OF PREPARATION OF BORAZINE

Method I : Borazine was originally prepared by Stock and Pohland in 1921 by the reaction of ammonia with diborane. The adduct $(BH_3.NH_3)_2$ is first formed which is then decomposed by heating in a closed tube at 200°C.

Cyclisation occurs to give borazine but yield is low because of the simultaneous formation of solid polymeric by-products.

$$3B_2H_6 + 6NH_3 \rightarrow 3(BH_3.NH_3)_2 \xrightarrow{200°C} 2B_3N_3H_6 + 12H_2$$

Method II : B-trichloroborazine, formed by the reaction between boron trichloride and ammonium chloride, can serve as an intermediate in the preparation of borazine. Small quantities are conveniently made by heating dry, powdered ammonium chloride in one end of a pyrex combustison tube at 170°C while a slow stream of boron trichloride vapour and nitrogen is passed over the powder.

B-trichlorobenzene sublimes as it is formed and condenses as needle-like crystals on the cool parts of the tube. Yields upto 35 percent have been reported. It is readily purified by vacuum sublimation. B-Trichlorobenzene is then reduced to borazine by sodium tetrahydroborate.

$$3NH_4Cl + 3BCl_3 \rightarrow 9HCl + Cl_3B_3N_3H_3$$

$$Cl_3B_3N_3H_3 + 3NaBH_4 \rightarrow B_3N_3H_6 + 3/2\ B_2H_6 + 3NaCl$$

As the equation shows, 3/2 moles of diborane are liberated for every mole of $B_3N_3H_3Cl_3$ used. This necessitates special precautions and the reaction is difficult to exploit on a large scale unless immediate use can he made of the diborane and suitable storage facilities are available.

An alternative procedure is to prevent the evolution of diborane by the addition of a tertiary aliphatic amine of relatively high boiling point. This reacts rapidly and completely with diborane to give the solid amine borane adduct.

Borazine is distilled in vacuo and purified by fractionation.

Method III : Borazine is also prepared by heating a mixture of lithium borohydride and ammonium chloride in vacuum at 230°C.

$$3NH_4Cl + 3LiBH_4 \xrightarrow{230°C} B_3N_3H_6 + 3LiCl + 9H_2$$

This reaction is the basis of a convenient laboratory method for borazine. Lithium borohydride, dissolved in diglyme, is used as the reducing agent. The yield of this method is 80 percent.

PROPERTIES OF BORAZINE

Physical Properties

Borazine is a colourless liquid having boiling point 64.5°C. Borazine decomposes slowly on storage even at 80°C, and small quantities of a white solid are deposited over a period of several days.

At room temperature, decomposition is much more rapid and hydrogen diborane and other volatile products are formed.

Borazine is isoelectronic which benzene. In physical properties, borazine is indeed a close analogue of benzene. The similarity of the physical properties of the alkyl substituted derivatives is more remarkable.

The ratio of the absolute boiling points of the substituted borazines to similarly substituted benzene derivatives is 0.93 ± 0.01.

This similarity of physical properties has unfortunately been overemphasized and led to a description of borazine as inorganic benzene. However, the properties of borazine and benzene are quite different.

Chemical properties : Main chemical properties of borazine are outlined below which are quite different from benzene.

1. Borazine readily undergoes addition reactions whereas benzene does not.

$$B_3N_3H_6 + 3HCl \rightarrow (H_2N-BHCl)_3$$

+HCl → No reaction

The hydrogen chloride derivative of borazine, if heated to 50-100°C, loses three molecules of hydrogen to give B-trichloroborazole.

$$(H_2N-BHCl)_3 \xrightarrow{-3H_3} B_3N_3H_3Cl_3$$

2. With water, borazine very slowly hydrolyses to hydrogen, boric acid and ammonia; this reaction is favoured by increase of temperature.

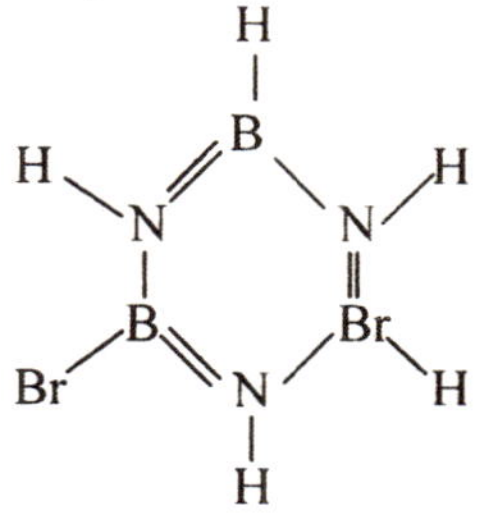

$$\text{(borazine)} + 3H_2O \rightarrow 2B(OH)_3 + 2NH_3 + 3H_2$$

On the other hand, benzene does not undergo hydrolysis.

3. Although, it was once believed that one mole of borazine forms an adduct with two moles of bromine and that this decompcses to B–2, 4-dibromobenzene as the main product, these reactions have not been repeated.

It appears that borazine reacts with an excess of bromine to give an orange solid, B-hexabromo-cyclotriborazine, $B_3N_3H_6Br_6$, and this decomposes on heating to give B-tribromoborazine and HBr.

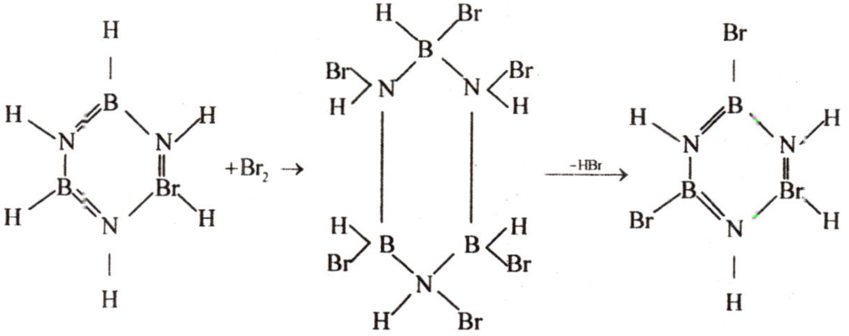

$$\text{(borazine)} + Br_2 \rightarrow \text{(adduct)} \xrightarrow{-HBr} \text{(bromoborazine)}$$

On the other hand, benzene with bromine in the presence of ferric bromide yields monobromoben

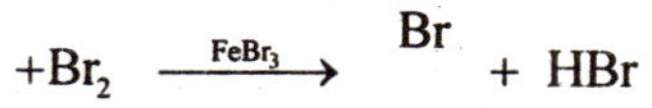

$$+Br_2 \xrightarrow{FeBr_3} \text{Br} + HBr$$

4. Benzene may be hydrogenated to produce the saturated compound cyclohexane. Hydrogenation of borazine results in polymeric materials of indefinite composition.

5. Borazine forms an adduct with methanol in which borazine : methanol = I : 3. This pyrolyses with the elimination of hydrogen to give B-trimethoxyborazine.

6. Borazine reacts with phenylmagnesium bromide to give B- arylated compounds. The main products are B-mono-and B-diphenylborazine together with a small amount of B-triphenyl-borazine.

7. Borazine undergoes a strongly exothermic reaction with anniline to produce, triaminoborine (III).

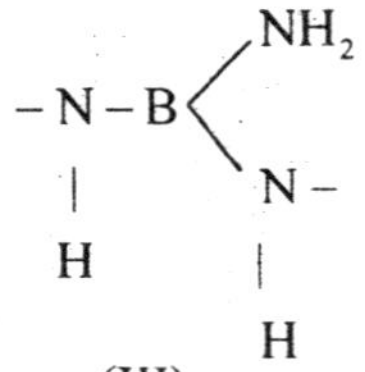

$$-\underset{\mathrm{H}}{\mathrm{N}}-\mathrm{B}\left\langle\begin{matrix}\mathrm{NH_2}\\ \underset{\mathrm{H}}{\mathrm{N}}-\end{matrix}\right.$$

(III)

8. When borazine is pyrolysed at temperature above 340°C the products include $B_6N_6H_{10}$ (I) and $B_5N_5H_8$ (II).

   These are the boron-nitrogen analogous of diphenyl and naphthalene respectively. B-2, 4-diaminoborazine is also formed, presumably after a break down of some of the borazine molecules. The thermal decomposition of borazine is not completely analogous to that of benzene which gives biphenyl, terphenyl and triphenylene but no naphthalene.

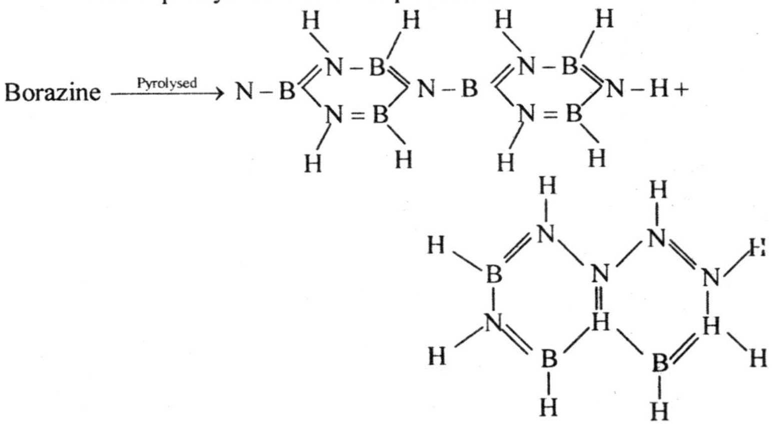

Borazine $\xrightarrow{\text{Pyrolysed}}$ N – B(N(H)–B(H)=N–B(H)=N(H)) N – B (N(H)–B(H)=N–B(H)=N(H)) N – H +

[Structure: fused B–N ring system with H substituents]

Fig. 1

STRUCTURE OF BORAZINE

X-ray crystallography has shown that boron and nitrogen atoms in borazine are arranged alternately in a planar, hexagonal ring, in which the B–N bond length is 0.147 ± 0.007nm.

Electron diffraction photographs of borazine and benzene are very similar and confirm that borazine has a benzene-like configuration.

The B–N bond length determined by electron diffraction is 0.44 ± 0.002 nm. These facts are in accordance with the hexagonal ring structure.

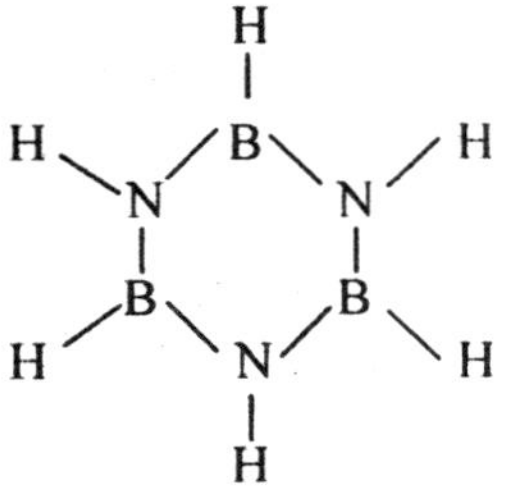

(a)

The bonding in borazine has been described in terms of resonance between the single-bonded triaminotriborine ring (a) and the Kekule structures (b) and (c) in which double bonds are present due to the participation of the lone pairs on nitrogen in the boron-nitrogen bonds. This results in a formal negative charge on the boron and formal positive charge on the nitrogen atom.

This resonance hybrid description is based on the similarities which exist between borazine and benzene. In borazine, the B–N bond length is 0.144 nm. which is very close to the value in benzene for C–C ad 1.42 nm.

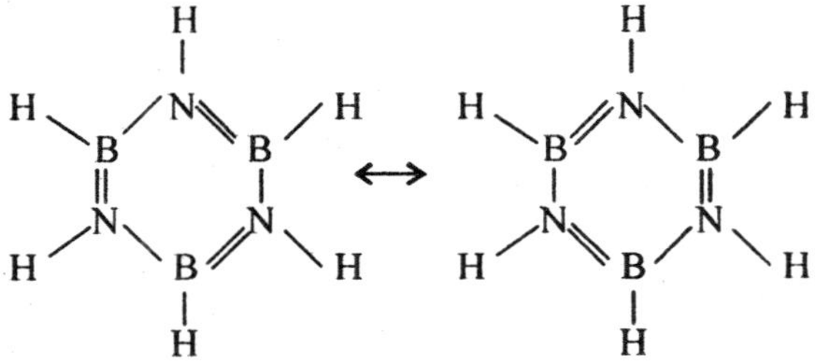

(b) ⟷ (c)

Fig. 2

The molecular orbital description of borazine takes into consideration the formation of σ and π-bonds in the ring. Because, of the difference in electronegativity between boron and nitrogen, the π-cloud in borazine is lumpy with more electron density localised on the nitrogen than boron. This partial localisation weakens the π-bonding in the ring. In addition, nitrogen retains some of its basicity and the boron some of the acidity. Polar species such as HCl can therefore attack the double bond between nitrogen and boron (see chemical reactions).

The different electronegativities of boron and nitrogen tend to stabilise bonding to boron by electropositive substituents and to nitrogen by electropositive substituents. Thus, in contrast to benzene, borazine readily undergoes addition reactions. If a criterion of aromaticity is taken to be the extent of delocalisation of electrons in π-orbitals, it appears that the aromatic nature of borazine is significantly less than that of benzene.

SUBSTITUTED BORAZINE

Main substituted borazines are discussed below

(a) **B-trimethyl borazine :** It is prepared by heating $(CH_3)_3B$ with ammonia at 320-340° at 20 atm for 2 hours. There are a number of stages to the reaction and the intermediate products can be isolated. However, the final product is B-trimethylborazine.

$$B(CH_3)_3 + NH_3 \rightarrow [(CH_3)_3B \leftarrow NH_3] \rightarrow (CH_3)_2B \overset{\pi}{-} NH_2 + CH_4$$

$$\downarrow$$

$$H_3B_3N_3(CH_3)_3 \leftarrow (CH_3)B \overset{2\pi}{-} NH + CH_4$$

At room temperature this compound (I) consists of colourless, volatile crystals. M. P. 31.8° B. P. 127°.

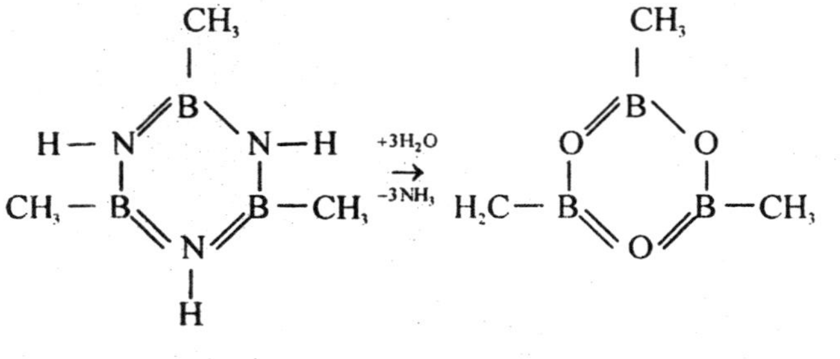

CH₃-substituted borazine ring (B–CH₃, N–H) $\xrightarrow[-3NH_3]{+3H_2O}$ CH₃-substituted boroxine ring (B–CH₃, O)

```
         CH3                                   CH3
          |                                     |
          B                                     B
        //  \                                 //  \
H — N        N — H     +3H2O                 O      O
    |        |          →                    |      |
CH3 — B       B — CH3  -3NH3     H3C — B          B — CH3
        \\  //                                \\  //
          N                                     O
          |
          H
```

It is stable in vacuum to 350°C. It is sensitive to moisture but is insoluble in water. It is soluble in many organic solvents. At 100°C water replaces the NH groups by oxygen and gives B-trimethyl boroxine (II).

(b) **Boroxine :** Isoelectronic with the borazine is boroxine, $H_3B_3O_3$(II). It is produced by the explosive oxidation of B_2H_6 or B_5H_9. The structure has been determined and found to be perfectly planar. Boroxine is even less stable and presumably had less n delocalisation than borazine, decomposing at room temperature to diborane and boron oxide.

In boroxine the B–O bond distance is 0.138 nm. The characteristic Raman frequency of the ring is 807 cm^{-1}. Boroxine exhibits aromatic properties of benzene.

Note : By looking at the structure, II, one can immediately say that this structure is not correct. But it is correct. The double bond between boron and oxygen is due to the participation of the lone pairs on oxygen. This result in a formal negative charge on the boron and a formal positive charge on the oxygen atom.

(c) **N-trimethyl Borazine :** It is obtained in 90 percent yield by heating a mixture of B_2H_6 and CH_3NH_2 in the correct proportions and at 180-200°C for 2 hours.

$$B_2H_6 + 2CH_3NH_2 \rightarrow 2[H_3B \leftarrow NH_2CH_3] \rightarrow 2[H_2B \overset{\pi}{-} NHCH_3] + 2H_2$$

$$\downarrow$$

$$2[HB \overset{2\pi}{-} NCH_3] + 2H_2$$

The final product polymerises to the trimer which is N-trimethyl borazine.

$$3\left(HB \overset{2\pi}{-} NCH_3\right) \rightarrow N-\text{trimethylborazine}$$

N-trimethyl borazine may also be prepared by reducing monomethyl ammonium chloride with lithium borohydride.

$$3CH_3NH_3Cl + 3\ LiBH_4 \rightarrow B_3N_3H_3(CH_3)_3 + 3LiCl + 9H_2$$

N-trimethyl borazine is a colourless mobile liquid, M. P. –90°C, B. P. 132°C. It is stable upto 300°C. With water at 60° the B–H link (III) is attacked and product is (IV).

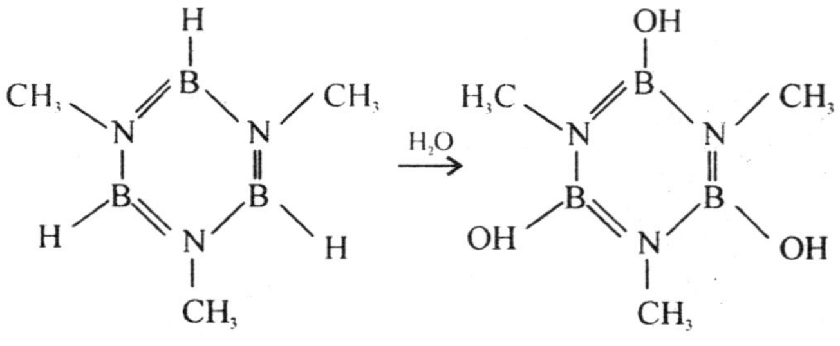

(III) $\xrightarrow{H_2O}$ (IV)

INTERHALOGEN COMPOUNDS

Introduction

The difference in electronegativity values among the elements of a halogen family gives rise to the formation of binary compounds. These are collectively acknowledged ad interhalogen compounds. The structures of these compounds are known with the varying degree of accuracy. The binary compounds formed by halogens amongst themselves are known as "Interhalogen Compounds." These have the general formula AX_n where n = 1, 3, 5 and 7.

There are never more than two different halogens in any known interhalogen compound. It means that ternary compounds of halogens with one another are not known. The probable reason for this is that such a complex molecule might be unstable which would be broken down to simpler compounds.

Selig (1964) of course claimed to have obtained a compound $BrClF_6$ due to the interaction between BrF_3 and chlorine which is a colourless solid, conducting due to self-ionisation of this compound to give rise ClF_2^+ and BrF_4^- ions. This needs further investigation. However, polyhalides containing three different halogens such as MIBeF and MIClBr and $MIFCl_3$ are known.

It will be noted that interhalogen compounds are of the type AX_n, when n is an odd number and X is always the lighter halogen when n is greater than unity. Because n is always odd, it follows that all the interhalogen compounds are diamagnetic having all velency electrons present either as shared (bonding) or unshared Pairs. It is expected that nay further interhalogens, if discovered, will contain an even number of atoms.

CLASSIFICATION OF INTERHALOGEN COMPOUNDS

The interhalogen compounds may be divided into four types on the basis of their formulae :

AX	*AX₃*	*AX₅*	*AX₇*
ClF			
BrF			
BrCl	ClF_3		
ICl	BrF_3	BrF_5	
IBr	ICl_3	IF_5	IF_7

where A and X are two different halogens.

With the exception of BrCl, ICl, IBr and ICl_3, all the remaining interhalogen compounds are halogen flourides. Greater the electro-negativity difference, $(X_F - X_A)$ (a may be I or Br), greater the number of fluorides given by atom A.

For example iodine $(X_F - X_I = 4.0 - 2.5 = 1.5)$ yields four fluorides viz. IF, IF_3, IF_5, and IF_7 while Br atom $(X_F - XBr = 4.0 - 2.8 = 1.2)$ yields only three fluorides namely BrF, BrF_3 and Br_5. The formation of three fluorides by Cl-atom $(x_F - X_{Cl} = 4.0 - 3.0 = 1.0)$ is taken to be an exception. Actually Cl– atom should yield less than three fluorides.

NAMING OF INTERHALOGEN COMPOUNDS

In naming interhalogen compounds, they are regarded as the halides of the more electropositive element.

Thus, ClF is named as chlorine monofluoride and not as fluorine monofluoride because the electronegativity of Cl atom is less than that of F atom, ICl_3 is named as iodine trichloride.

CAUSES FOR THE FORMATION OF INTERHALOGEN COMPOUNDS

1. In the case of halogens, the electronegative behaviour of elements decreases with rise in size of atom and atomic number. So iodine shows basic properties and due to this factor it combines with other halogens to form interhalogen compounds.
2. Small size and strong oxidising action of fluorine result in formation of ClF, BrF_3, IF_7, etc.

3. The maximum number of covalently bonded atoms in these compounds increases with the size of oxidised atom of halogen. The value of fluorine for chlorine is 3 while for bromine is 5 but for iodine it is 7.
4. AX_1 and AX_3 types of interhalogen compounds are formed when electronegative difference is not too great, *e.g.*, ClF, BrF, IBr, ClF_3 etc.
5. AX_5 and AX_7 types of interhalogen compounds are formed by atoms such as bromine and iodine which combine with small size halogen atoms such as fluorine, since it is possible to associate more small atoms around a large central atom.
6. As the smaller halogen atoms are grouped around the large one and the number of smaller atoms of halogen are around the large atom of halogen, this is guided by the radius ratio, *i.e.*,

$$\text{Radius ratio} = \frac{\text{Radius of larger halogen}}{\text{Radius of smaller halogen}}$$

EXCEPTIONS TO THE FORMATION OF HIGHER INTERHALOGENS

All the interhalogens are of the type AX_n where n is an odd number. As n is always odd all interhalogen compounds are less stable and it is therefore expected that any further interhalogens, if discovered, must contain an even number of atoms. No inter-halogen compound containing even number of atoms could be prepared so far

GENERAL METHODS OF PREPARATION OF INTERHALOGENS

Following have been the important general methods by which interhalogen compounds can be prepared.

1. By the Direct Combination of Halogens

All the interhalogens, except IF_7, could be prepared by the direct combination of the halogens under appropriate conditions. The following examples are notworthy. For the preparation of AB_3 and AB_5 interhalogen compounds excess of B atom should be used in the reaction mixture.

$$Cl_2 + F_2 \text{ (equal volumes)} \xrightarrow{250^o} 2ClF$$

$$Cl_2 + 3F_2\ (\text{excess}) \xrightarrow{250^{\circ}} 2ClF_3$$

$$I_2 + Cl_2\ (\text{liquid})\ (\text{in equimolecular amount}) \rightarrow 2ICl$$

$$I_2 + 3Cl_2\ (\text{liquid and in excess}) \rightarrow 2ICl_3$$

$$Br_2\ (\text{vapour}) + 3F_2\ (\text{diluted with } N_2) \rightarrow 2BrF_3$$

$$I_2 + 3F_2\ (\text{in freon}) \xrightarrow{-78^{\circ}C} 2ClF_3$$

$$Br_2 + 5F_2\ (\text{excess}) \rightarrow 2BrF_5$$

2. From Lower Interhalogens

It is possible to prepare higher interhalogens by the interaction of lower interhalogens with halogens. This method is particularly used for the preparation of halogen fluorides. The following examples demonstrate the method.

$$ClF + F_2 \rightarrow ClF_3$$

$$ClF_3 + F_2\ (\text{excess}) \xrightarrow{350^{\circ}} ClF_5$$

$$BrF_3 + F_2 \xrightarrow{200^{\circ}} BrF_5$$

$$IF_5 + F_2 \xrightarrow{270^{\circ}} IF_7$$

3. Miscellaneous Methods

Some miscellaneous methods have been given by the following examples :

$$6HCl + KIO^3 + 2KI \rightarrow 2KCl + 3H^2O + 3ICl$$

$$Cl_2 + ClF_3 \xrightarrow{250-350^{\circ}} 3ClF$$

$$KCl + 3F_2 \xrightarrow{200^{\circ}} KF + ClF_5$$

$$3I_2 + 5AgF \rightarrow 5AgI + IF_5$$

$$8F_2 + PbI_2 \rightarrow PbF_2 + 2IF_7$$

CHEMICAL PROPERTIES OF INTERHALOGENS

(i) Generally interhalogen compounds are more reactive than the halogens except fluorine because the A – X bond is weaker than the X – X bond in halogens. However, their chemical properties are similar to those of halogens.

(ii) Interhalogen compounds behave as oxidising agents and attack most of the elements resulting in a mixture of halides.

(iii) These are hydrolysed to form halide and oxyhalide ions. The oxyhalide is generally formed from the larger halogen present. Oxidation state of larger halogen does not alter during hydroysis.

$$BeF_5 + 3H_2O \rightarrow 5F^- + BrO^- + 6H^+$$

$$ICl + H_2O \rightarrow Cl^- + IO^- + 2H^+$$

(iv) Strength of A – X bond : Now we know that greater the separation of two elements on the electronegativity series, the greater the strength of bond between them. Thus, out of halogens fluorine and iodine differ most, the combination between them is most vigorous and consequently the stability of their compounds is also greater.

GENERAL PROPERTIES OF INTERHALOGENS

1. **State :** In general, interhalogen compounds are covalent gases (e,g., ClF, BrF, ClF_3, IF_7) due to the small electronegativity difference, liquids (*e.g.*, BrF_3, BrF_5, and IF_5) or solids (*e.g.*, ICl, IBr, ICl_3, IBr_3). The melting and boiling points increase as the difference in electronegativity increases.

2. **Colour :** Many of the interhalogen compounds containing fluorine and colourless. However, those made up of heavier halogens are coloured. The colour becomes deeper with the increase of the molecular weight of the compound. In this behaviour these compounds resemble the halogens themselves.

3. **Diamagnetic nature :** From The structures of nterhalogens it can be seen that all the valence electrons in interhalogen compounds are present as bonding (*i.e.*, lon pairs) electron pairs. Therefore, these compounds are diamagnetic in nature.

4. **Heats of formation :** All these compounds are having low heats of formation. For example, these values for IBr and BrCl are – 10.5 and 14.5 kJ/mole respectively. These are extremely volatile compounds. In general, compounds having fluorine are more volatile than those containing chlorine, bromine or iodine.

5. **Boiling points :** For each type of interhalogen compounds, the boiling points tend to increase with the increase in the electronegativity deference between A and B atoms.

6. **Thermal stability :** Thermal stability of AB type inter-halogen compounds gets decreased with the decrease in electronegativity difference between A and B atoms. Hence, the order of stability of some AB compounds ins as fallows :

   IF (= 1.5) > BrF (1.2) > ClF (1.0) > ICl (0.5) > IBr (0.3) > BrCl (0.2)

   In parenthsese the electronegativity difference between A and B atoms have been indicated. The above order could also be explained by saying that greater the difference between the electronegativity values of A and B, the more polar the A – B bond and hence greater the thermal stability of AB compound.

7. **Reactivity :** AB type compounds have been found to be more reactive than A_2 and B_2 molecules, because A – B bond in AB compounds is weaker that A – A and B – B bonds in A_2 and B_2 molecule respectively. Hence AB type compounds convert the metals into a mixture of two halides. For example :

   $$ICl + 2Na \rightarrow NaI + NaCl$$

   The order of reactivity of some interhalogen compounds has been found to be as $ClF_3 > BrF_3 > IF_7 > BrF_5 > IF_5 > BrF$.

8. **Hydrolysis :** Hydrolysis of interhalogens yields halogen acid and oxy-halogen acid. The oxyhalogen acid is of larger (*i.e.*, central) halogen atom. Examples are as follows :

$BrCl + H_2O \rightarrow HCl$ (Halogen acid) $+ HOBr$ (Oxy-halogen acid)

$ICl + H_2O \rightarrow HCl + HIO$

$ICl_3 + 2H_2O \rightarrow 3HCl + HIO_2$

$IF_5 + 3H_2O \rightarrow 5HF + HIO_3$

$BrF_5 + 3H_2O \rightarrow 5HF + HBrO_3$

$IF_7 + 6H_2O \rightarrow 7HF + H_5IO_6$

   It is to be noted that the oxidation state of A atom does not change during hydrolysis.

9. **Reaction with non-metallic and metallic oxides :** NOen-metallic and metallic oxides are fluorinated by interhalogens having fluorine to yield the corresponding fluorides.

   $$4BrF_3 + 3SiO_2 \rightarrow 3SiF_4 + 2Br_2 + 3O_2$$

   $$4BrF_3 + 2WO_3 \rightarrow 2WF_6 + 2Br_2 + 3O_2$$

10. **Addition reactions :** The diatomic interhalogens (AB type compounds) add at olefinic double bond sites.

$$- CH = CH - + ICl \rightarrow - CHI - CHCl -$$

$$CH_2 = CH_2 + BrCl \rightarrow CH_2Br - CH_2Cl$$

11. **Reaction with alkali metal halides :** Reaction with alkali meta! halides yields polyhalides like $NaIBr_2$, CsIBrCl, $KICl_4$ etc.

$$IBr + NaBr \rightarrow NaIBr_2;\ ICl + KCl \rightarrow KICl_2$$

$$BrCl + CsI \rightarrow CsLBrCl;\ ICl_3 + KCl \rightarrow KICl_4$$

12. **Self-ionisation and acid-base (neutralisation) reactions :** From the available evidence it is evident that the liquid interhalogen compounds undergo self-ionisation which involves the halide ion transfer. Following examples demonstrate how the self-ionisation of liquid interhalogen compounds occurs.

(i) *Liquid ICl :* The self-ionisation of liquid ICl can be represented in the form of conjugate acid-base pair as follows :

$$\underset{(Acid_1)}{ICl} + \underset{(Base_2)}{ICl} \rightleftarrows \underset{(Acid_2)}{I^+} + \underset{(Base_1)}{ICl_2^-}$$

Hence, the compounds which produce I^+ ions in liquid ICl act as solve-acids in this solvent while those which produce ICl_2^- ions act as solvo-bases.

(ii) *Liquid IBr :* The self-ionisation of liquid IBr is represented as :

$$\underset{(Acid_1)}{IBr} + \underset{(Base_2)}{IBr} \rightleftarrows \underset{(Acid_2)}{I^+} + \underset{(Base_1)}{IBr_2^-}$$

(iii) *Liquid BrF_3 :* The self-ionisation of liquid BrF_3 can be represented as:

$$BrF_3\ (Acid_1) + BrF_3\ (Base_2) ** BrF_2^+ (Acid_2) + BrF_4^- (Base_1)$$

(iv) *Liquid IF_5 :* The self-ionisation of liquid IF_5 can be represented as :

$$IF_5\ (Acid_1) + IF_5\ (Base_2) \rightleftarrows IF_4^+ (Acid_2) + IF_6^- (Base_1)$$

13. **Lewis acid strength :** Lewis acid strength of some AB type interhalogens gets decreased in the order : (Halogens have also

been included in the series for the sake of comparison) ICl > > BrCl > IBr > I_2 > Br_2 > Cl_2

14. **Halogenation reactions :** BrF_3, IF_5 and ICl have been reported to be good halogenating agents.

For example :

(a) When Au and Ag are dissolved in liquid BrF_3 in correct molar proportions, these metals get converted into AuF3 and AgF respectively (Fluorination). These salts react together to yield the fluorinated complex, $Ag^+ [AuF_4]^-$.

$$Ag \xrightarrow{\text{Liq. } BrF_3} AgF; \; Au \xrightarrow{\text{Liq. } BrF_3} AuF_3$$

$$AgF + AuF_3 \rightarrow Ag^+ [AuF_4]^-$$

(b) Sulphur is converted into SF_4 by IF_5 (Fluorination)

$$IF_5 + S \xrightarrow[100-300^0]{12 \text{ hours}} SF_4 + IF$$

(c) C_6H_6 is converted into C_6H_5I by ICl (Iodination).

15. **Oxidising power :** The oxidation power of the fluorides of bromine (BrF, BrF_3 and BrF_5) gets increased with the increase of the oxidation number of Br atom in these flourides. Hence the oxidising power of these fluorides has been in the order : $Br^+F < Br^{3+} F_3 < Br^{5+} F_5$.

Oxidising power of ClF_3, BrF_3 and IF_5 has been in the order :

$$ClF_3 > BrF_3 > IF_5$$

Structure– All interhalogen compounds are diamagnetic having al valency electrons present either as shared or unshared pairs.

The structure of each of the interhalogens is in accordance with the number of election pairs in the valency shell of the central atom. For example:

(i) ClF_3 has trigonal bipyramidal disposition because central chlorine atom has three unpaired electrons

(ii) IF_5 has octahedral structure with one position occupied by lone pair because central iodine atom has five unpaired electrons.

Let us consider the different interhalogens one by one.

COMPOUNDS OF THE AX TYPE

General Characteristics of Compounds of AX Type

(i) These compounds can be prepared by the direct union of two halogens under suitable conditions. For example,

$$Cl_2 + F_2 \xrightarrow{250^\circ C} 2ClF$$

$$Br_2 + F_2 \xrightarrow{60^\circ C} 2BrF$$

$$I_2 + Cl_2 \xrightarrow{35^\circ C} 2ICl$$

(ii) Interhalogen compounds of the AX type resemble the halogens in many physical properties. In general, compounds containing fluorine are more volatile than those in which it is replaced by chlorine. and so on down the group.

(iii) The more polar the bond, the more stable the molecule. Thus, the stabilities of these compounds are in the following order :

$$IF > BrF > ClF > ICl > IBr > BrCl$$

(iv) Compounds of AX type are more reactive than the elements X because the A – X bond energy is less than the bond energy of hlogen X – X.

(v) All the compounds of AX type undergo hydrolysis and form addition compounds with olefins.

$$AX + 2H_2O \rightarrow HOA + H_3O^+ + X^-$$

$$\begin{matrix} CH_2 & & A & & CH_2A \\ \| & + & | & \rightarrow & | \\ CH_2 & & X & & CH_2X \end{matrix}$$

Let us now discus the different examples of this type of interhalogens one by one.

1. Chlorine Monofluoride ClF

Preparation

(i) It may be prepared by the direct union of two compounds at 250°C through a copper tube.

$$Cl_2 + F_2 \xrightarrow{250^\circ C} 2ClF$$

(ii) It is prepared by the reaction of chlorine with hydro-fluoric acid at liquid air temperature.

$$HF + Cl_2 \rightarrow HCl + ClF$$

(iii) Another convenient method is by mixing ClF_3 and Cl_2 in equimolecular quantities.

$$ClF_3 + Cl_2 \rightarrow 3ClF$$

Properties

(i) It is an almost colourless gas having a characteristic odour.

(ii) It reacts like fluorine but more readily. It may be probably due to its easy dissociation to from fluorine atoms. With carbon monoxide, it forms carbonyl chlorofluoride.

$$ClF + CO \rightarrow COClF$$

It reacts with sodium chloride, forming sodium fluoride and chlorine.

$$NaCl + ClF \rightarrow NaF + Cl_2$$

With silica, it forms silicon tetrafluoride.

$$4ClF + SiO_2 \rightarrow SiF_4 + 5Cl_2O$$

(iii) It inflames cotton wool and filter paper in the same way as F_2.

2. Bromine Monofluoride : BrF

Preparation

It may be prepared by saturating gaseous bromine with fluorine at 50°C. or by the action of BrF_3 on bromine.

$$Br_2 + F_2 \rightarrow 2BrF$$

$$BrF_3 + Br_2 \rightarrow 3BrF$$

Properties

(i) It is pale brown gas which condenses to a dark red liquid and freezes to a yellow crystalline solid.

$$BP = 23°C$$

$$MP = -34°C$$

(ii) It is reactive and undergoes disproportionation.

$$3BrF \rightarrow BrF_3 + Br_2$$

A small quantity of BrF_5 is also produced.

(iii) on heating, it decomposes into elements.

3. Bromine Monochloride BrCl

Preparation

It may be prepared by passing chlorine gas into liquid bromine below 10°C.

$$Br_2 + Cl_2 \rightleftharpoons 2BrCl$$

Properties

(i) It is highly volatile reddish-brown mobile liquid. Its vapours are irritating to eyes.

(ii) It is soluble in water, ether and carbon disulphide.

(iii) When heated to 10°C, it decomposes with the evolution of chlorine.

4. Iodine Monochloride, ICl

This compound was discovered by Davy and Gay Lussac in 1814.

Preparation

(i) It is prepared by adding iodine to liquid chlorine; the mixture is kept at 35°C for 24 hours to ensure complete reaction.

$$I_2 + Cl_2 \rightarrow 2ICl$$

It is obtained as a dark liquid which solidifies on long standing at room temperature

(ii) In aqueous solution, it canbe obtained by the action of hydrochloric acid on a mixture of postassium iodide and potassium iodate.

$$KIO_3 + 5KI + 6HCl \rightarrow 6KCl + 3H_2O + 3I_2$$

$$KIO_3 + 2I_2 + 6HCl \rightarrow KCl + 5ICl + 3H_2O$$

The above reaction is quantitative which involves two stages: These stages can be observed by the appearance and disappearance of the violet colour of the free iodine.

Properties

(a) Iodine monochloride exists in the two forms :

(i) Stable form : It has red needle-like crystals melting at 27.3°C:

(ii) Metastable form : It is a black solid melting at 14°C.

(b) It is readily hydrolysed by water forming hypo-iodite and chloride ions.

$$ICl + H_2O \rightarrow Cl^- + OI^- + 2H^+$$

The hydrolysis can be checked by adding n/5 HCl acid

(c) When electricity is passed through iodine monochloride, iodine is liberated at the cathode and chlorine at the anode, indicating that the compound is a chloride of chloride of monovalent iodine.

$$2ICl \rightarrow I^+ + [ICl_2]^-$$

$$2I^+ + 2e^- \rightarrow I_2 \text{ (At cathode)}$$

$$2ICl_2^- \rightarrow I_2 + 2Cl_2 + 2e \text{ (At anode)}$$

Thus, liquid iodine monochloride is used as an ionising solvent in which the heavier alkali-metal chlorides KCl, RbCl, CsCl, etc, dissolve.

(d) It is used as a Wij's reagent in the estimation of the iodine number, a measure of unsaturation in fats and oils.

$$-CH=CH- + I-Cl \rightarrow -\underset{\displaystyle I}{\underset{|}{CH}}-\underset{\displaystyle Cl}{\underset{|}{CH}}-$$

(e) When treated with organic compounds, it is used to iodinate them.

$$\text{Salicylic acid} + ICl \xrightarrow{\text{In nitrobenzene}} 3:5\text{di}-\text{iodo salicylic acid}$$

(f) ICl forms charge-transfer type complexes with organic bases like pyridine (I) dioxane (II)

Uses

(i) A solution of ICl in glacial acetic acid finds use for the determination of iodine number which measures the unsaturation of oils and fats.

(ii) It also finds use to iodate and chlorinate organic compounds.

(iii) The solution of Icl finds use as a catalyst in oxidinsing As_2O_3 by $Ce(SO_4)_2$.

(iv) It also finds use to prepare poly-halides

(v) Because of the formation of I+ cation, ICL also finds use as an electrophilic iodinating agent. For example, it converts salicylic acid into 3, 5-diodosalicylic acid in nitrobenzene.

5. Iodine Monobromide IF

Preparation

It is obtained by the direct combination of bromine with iodine.

$$I_2 + Br_2 \rightarrow 2IBr$$

Properties

(i) It crystallises at 42oC to a black solid which boils at 116oC to give red vapours. Iodine monobromide undergoes decomposition when heated.

$$2IBr \rightleftarrows I_2 + Br_2$$

(ii) When it dissolves in non-solvating, it forms red solution whereas it forms yellow solution in solvating solvents,

(iii) IBr is an electrical conductor when in the molten state. It ionises as :

$$IBr \rightarrow I^+ + Br^-$$

$$2I^+ + 2e^- \rightarrow I_2 \text{ (At cathode)}$$

Thus in IBr, iodine is present as I^+.

6. Iodine Monofluoride If

This compound has never been isolated. Evidence for the existence of IF was reported by Durie in 1960 and Bhagi (1989) from the spectrum of CH_3I after burning in fluorine. This spectrum is similar to that arising when fluorine gas is made to strike on the surface of iodine crystals.

INTERHALOGENS OF THE TYPE AX_3

Three compounds of this series are known. These are :

(i) Chlorine Trifluoride, ClF_3

(ii) Bromine Trifluoride, BrF_3

(iii) Iodine Trichloride, ICl_3,

Of the AX_3 compounds. ClF_3 is most reactive but; BrF_3 is more useful in th preparative work. Thermal stabilities of the type AX_3 compounds fall in the following order.

$$BrF_3 > ClF_3 > ICl_3$$

Let us discuss the interhalogens of type AX_3 :

1. Chlorine Trifluoride ClF_3

Preparation

It may be prepared by passing either chlorine or chlorine monofluoride mixed with an excess of fluorine through a nickel tube maintained at 220°C.

$$F_2 + Cl_2 \rightarrow 2ClF$$

$$ClF + F_2 \rightarrow ClF_3$$

The product ClF_3 is condensed by liquid air and is separated by distillation.

Properties

(i) It is a colourless gas which condenses to a green liquid (b.p = – 12°C) and freezes to a white solid.

(ii) It is more reactive than fluorine and thus attacks most elements with explosion. It attacks glass and inflames number of ograinc compounds.

(iii) Liquid ClF_3 reacts with water with a loud crack forming a red liquid which crystallises at – 70°C and has been found to contain ClFO.

Uses

(i) It is used for fluorinating a number of organic compounds, *e.g.*, polyhalides.

(ii) It is used in incendiary and in cutting oil well tubes.

(iii) It is used as an oxidiser of propellants.

2. Bromine Trifluoride BrF_3

Preparation

(i) It may be obtained by mixing bromine vapour and fluorine in a stream of nitrogen, or by the action of ClF_3 on bromine at 10°C

$$Br_2 + 3F_2 \rightarrow 2BrF_3$$

$$2ClF_3 + Br_2 \rightarrow 2BrF_3 + Cl_2$$

This compound can be purified by distillation.

(ii) It is also formed when BeF undergoes disproportionation.

$$3BrF \rightarrow BrF_3 + Br_2$$

Properties

(i) It is a colourless to green yellow fuming liquid having B.P. 127°C,

(ii) It ionises considerably.

$$2BrF_3 \rightleftarrows BrF_2^+ + BrF_4^-$$

Thus, substances producing BrF_2^+ ions are acids and BrF_4^- ions are bases in this solvent. Bromine trifluoride reacts with metals and metallic oxides to give compounds of which the following examples may be given as :

$$[BrF_2^+]\ \ [BrF_4^-]$$

$$[BrF_2^+]\ \ [SbF_6^-]$$

$$\left[BrF_2^+\right]_2 [SnF_6^{2-}]$$

(iii) It reacts rapidly with glass and silica material and sets fire to wood, paper, etc.

$$3SiO_2 + 4BrF_3 \rightarrow 2SiF_4 + 2SiF_4 + 2Br_2 + 3O_2$$

(iv) It reacts with Br_2 to form BrF

$$BrF_3 + Br_2 \rightarrow 3BrF$$

(v) Action on metal fluorides. It dissolves certain metal fluorides into it to form tetra fluoroborites, $M^+[BrF_4]^-$ or hexafluoroborates, $M^{3+}[BrF_6]^{3-}$. For example :

$KF + BrF_3 \rightarrow K^+[BrF_4]^-$ (potassium tetrafluoroborate)

$SbF_3 + BrF_3 \rightarrow Sb^{3+}[BrF_6]^{3-}$ (Antimony hexafluoroborate)

(vi) Action on metals, their oxides and chlorides. It converts metals, their oxides and chlorides into their fluorides. For example :

$$6M + 2BrF_3 \rightarrow 6MF + Br_2$$

$$3MCl + BrF_3 \rightarrow 3MF + BrCl_3$$

Uses

(i) Due to the fact that BrF_3 converts many metals, their oxides and chlorides into their fluorides, it finds use as a fluorinating agent. Many metal fluorides are also converted into complex fluorides.

(ii) BrF_3 is used to prepare polyhalides.

3. Iodine Trichloride ICl_3

Preparation

(i) It is prepared by heating I_2O_5 with hydrogen chloride

$$I_2O_5 + 10HCl \rightarrow 2ICl_3 + 5H_2O + 2Cl_2$$

(ii) It is prepared by passing chlorine over iodine until this is converted into ICl and then raising temperature to 100°C and increasing the stream of chlorine. When gaseous products are condensed. ICl_3 is obtained.

$$I_2 + Cl_2 \rightarrow 2ICl$$

$$ICl + Cl_2 \rightleftarrows ICl_3$$

(iii) It is best prepared by condensing chlorine on iodine at –80°C.

$$I_2 + 3Cl_2 \rightarrow 2ICl_3$$

(iv) It is also formed along with ICl by the action of HI on Cl_2.

$$4HI + 4Cl_2 \rightarrow 4HCl + I_2 + ICl + ICl_3$$

Properties

(i) Iodine trichloride consists of lemon yellow crystals which melt at 101°C to a reddish-brown liquid. It is soluble in organic solvents, liq. NH_3 and liq. SO_2.

(ii) With water, it undergoes hydrolysis.

$$2ICl_3 + 3H_2O \rightarrow ICl + 5HCl + HIO_3$$

(iii) It dissolves in organis solvents, liquid sulphur dioxide and liquid ammonia.

(iv) Iodine tri-chloride combines with metal chlorides to form addition compounds, $MICl_4$ These compounds dissociate on exposusre to air, giving ICl_3 and MCl.

$$MCl + ICl_3 \rightleftarrows MICl_4$$

(v) With fluorides of K, Rb, Cs it gives polyhalides of $MICl_3F$ type.

$$MF + ICl_3 \rightarrow MICl_3F$$

(iv) ICl_3 is completely hydrolysed by H_2O.

$$2ICl_3 + 3H_2O \rightarrow 5HCl + HIO_3 + ICl$$

(v) ICl_3 dissociates on heating at 68°C yielding ICl and Cl_2.

$$ICl_3 \rightarrow ICl + Cl_2$$

(iv) It reacts with CS_2 as :

$$4ICl_3 + 3CS_2 \rightarrow CCl_4 + 2CSCl_2 + 2S_2Cl_2 + 2I_2$$

(vii) Like ICl, ICl_3 also gets decomposed by KOH solution

$$3ICl_3 + 12KOH \rightarrow 9KCl + KI + 2KIO_3 + 6H_2O$$

(viii) Molten ICl_3 is having a high conductivity (= 8.4 × 10–3 ohm^{-1} cm^{-1}). 'The liberation of I2 and Cl2 at both electrodes reveals that ICl3 is ionised into $[I^{3+}Cl_2]^+$ and $[I^{3+}Cl_4]^-$ ions.

$$2ICl_3 \rightleftarrows [ICl_2]^+ + [ICl_4]^- \text{ (Ionisation)}$$

$$2[ICl_2]^+ 2e^- \rightarrow I_2 + 2Cl_2 \text{ (At cathode)}$$

$$2[ICl_4]^- 2e^- \rightarrow I_2 + 4Cl_2 \text{ (At anode)}$$

Uses

ICl_3 finds use in medicines. It also finds use in the preparation of polyhalides.

INTERHALOGENS OE THE TYPE AX_5

Two compounds of this series are known, namely, BrF_5 and IF_5. Out of these two compounds, BrF_5 is too reactive to be suitable for preparing fluorides.

1. Bromine Pentafluoride BrF_5

Preparation

(i) It is prepared by passing fluorine into bromine trifluoride at 100°C, and heating the gaseous product to 200°C.

$$F_2 + BrF_3 \rightarrow BrF_5$$

The product is separated by fractional distillation.

(ii) BrF_5 may be prepared by the action of fluorine (diluted with N_2) on the excess of bromine vapour in a copper vessel at 200°C.

$$Br_2 + 5F_2 \rightarrow 2BrF_5$$

Properties

(i) It is highly reactive and hence is generally diluted with nitrogen whenever required for the preparation of fluorides. It is a colourless fuming liquid, B.P. = 40.5°C.

(ii) Liquid bromine pentafluoride reacts explosively with water.

$$BrF_5 + H_2O \rightarrow 2HF + BrOF_3$$

(iii) When heated to 400 – 500°C, it dissociates into the elements.

$$2BrF_5 \rightarrow Br_2 + 5F_2$$

(iv) It reacts with most metals and non-metals, and with the chlorides, bromides and iodides of the alkali metals.

Uses

(i) It is used in organic synthesis.

(ii) It is used as an oxidiser for propellants.

(iii) It is used for the preparation of fluorides.

2. Iodine Pentafluoride IF_5

Preparation

(i) It is prepared by passing fluorine (diluted with nitrogen) over iodine in a heated quartz tube.

$$I_2 + 5F_2 \rightarrow 2IF_5$$

If any IF_7 is formed, it is easily removed as it is highly volatile.

$$IF_5 + F_2 \rightarrow IF_7$$

(ii) It is also prepared by the action of iodine on silver fluoride, as represented by the equation.

$$5AgF + 3I_2 \xrightarrow[\text{heat}]{\text{Red}} IF_5 + 5AgI$$

(iii) It may be prepared by the action of fluorine on heated iodine pentoxide.

$$2I_2O_5 + 10_2 \rightarrow 4IF_5 + 5O_2$$

Properties

(i) *It is a colourless liquid (B.P. = 97°C)* : Liquid IF_5 conducts electricity and self-ionises.

$$2IF_5 \rightarrow IF_4^+ + IF_6^-$$

It decomposes on heating to I_2 and IF_7.

$$7IF_5 \rightarrow I_2 + 5IF_7$$

(ii) *Reaction with water* : It fumes strongly in air and reacts rapidly with water to form HF and I_2O_5

$$2IF_5 + 5H_2O \rightarrow 10HF + I_2F_5$$

(iii) *Reaction with elements* : Metals like Ag, Hg, Cu, Fe, Mg and Cr are slowly attacked by IF_5. Sulphur, red phosphorus, silicon, bismuth, tungsten and arsenic react at once with IF_5, usually with incandescene.

(iv) *Reaction with oxides* : I_2O_5, P_2O_5 and V_2O_5 dissolve in hot IF_5, forming IOF_3, POF_3 and VOF_3 respectively.

$$3IF_5 + I_2O_5 \rightarrow 5IOF_3$$

$$2IOF_3 \xrightarrow{100^\circ C} IF_5 \text{ (black)} + IO_2F \text{ (white solid)}$$

I_2O_4 dissolves in IF_5 to give IOF_3 and some I_2.

The oxides P_2O_5, V_2O_5, Sb_2O_5, Cr_2O_3 and WO_3 dissolve in hot IF_5 to form POF_3, VOF_3, SbF_5. IO_2, CrO_2F_2 and WO_3.

$2IF_5$ respectively. Nitrogen (IV) oxide, NO_2, dissolves in IF_5 to give a cream coloured crystalline compound, $IF_5.NO_2$.

(v) *Reaction with carbonhalides* : With CCl_4, Iodine penta– fluoride yields CCl_3F, ICl and ICl_3.

$$CCl_4 + IF_5 \rightarrow CCl_3F + ICl + ICl_3$$

The reaction is not quantitative with CCl_4 but with CBr_4 it yields a mixture of bromofluoromethanes. With Cl_2, it yields ClF_3

(vi) Reaction with certain potassium salts : With potassium periodate, it yields $KIO_4.IF_5$.

$$KIO_4 + IF_5 \rightarrow KIO_4\ IF_5$$

With potassium fluoride, it yields KIF_6.

$$KF + IF_5 \rightarrow K[IF_6]$$

INTERHALOGENS OF THE TYPE AX_7

Iodine heptafluoride, IF_7, is the only interhalogen compound of this series. The compound is of considerable theoretical importance because iodine is the only example of a non-transition element using three d-orbitals for bonding.

Preparation

An impure sample of iodine heptafluoride is prepared by passing a mixture of vapours of IF_5 and fluorine through a platinum tube at 270°C.

$$IF_5 + F_2 \rightarrow IF_7$$

The issuing gases are condensed and IF_7 is purified by distillation. However, IF_7 so obtained is contaminated with IOF_5.

A pur sample of IF_7 is obtained by the fluorination of PdI_2.

$$PdI_2 + 8F_2 \rightarrow 2IF_7 + PbF_2$$

Properties

(i) It is colourless gas which canbe condensed to a colourless liquid or to a colourless crystalline solid.

(ii) Chemically, it is a comparable to chlorine trsifluoride or bromine pentafluoride in activity. Like them, it reacts with most of the metals except the platinum group and many non-metals including Cl_2 and I_2, many inorganic compounds and a variety of organic materials. When vapour of IF_7 is blown through water, periodate ions and fluoride ions are formed in solution.

$$IF_7 \xrightarrow{H_2O} \underset{\text{Per - iodiate ion}}{IO_4^-} + 7F^-$$

(iii) When heated with hydrogen, it explodes.

(iv) At 350°C in the presence of HgF_2, it fluorinates CCl_2F_2 to $CClF_3$.

IF_7 does not react with alkali metal fluorides to form MIF_8.

Uses

It is used as a fluorinating agent.

APPLICATIONS OF INTERHALOGEN COMPOUNDS

(i) Iodine monochloride is used in the estimation of iodine number which is a measure of the unsaturation in acids and fats.

$$-CH=CH- + ICl \rightarrow -\underset{\substack{| \\ I}}{HC}-\underset{\substack{| \\ Cl}}{CH}-$$

(ii) Iodine monochloride is used to iodinate organic compounds, though chlorination may occur depending upon the experimental conditions.

(iii) Liquid BrF_3 and ClF_3 have been used as fluorinating agents for the preparation of a number of metal fluorides.

(iv) BrF_3, ClF_3 and ICl have been used for the preparation of polyhalides.

$$NaBr + ICl \rightarrow NaBr.ICl$$

(v) Liquid BF_3 has been used to carry out various types of reactions, and new compounds such as $BrF_2.SbF_6$, $(BrF_2)_2SnF_6$ have been prepared from reactions in this solvent.

(iv) BrF_5 is used as an oxidiser for propellants.

STRUCTURES ND SHAPES OF THE COMPOUNDS AX

The electronic structures of type AX interhalogen compounds are

$$:\underset{\cdot\cdot}{\overset{\cdot\cdot}{A}}:\underset{\cdot\cdot}{\overset{\cdot\cdot}{X}}: \quad \text{or} \quad A - X$$

where A is more electropositive than X and shares an electron pair with X, forming single covalent linkage.

Table 3

Compound	*Structure*	*Bond Length* Å
Cl – F	:Cl:F:	1.628
Br – F	:Br:F:	1.756
Br – Cl	:Br:Cl:	2.138
I – C	:I:Cl:	2.321
I – Br	:I:Br:	2.470

On this basis the structures of different interhalogens are written as given in Table 3.

From the above table it evident that the difference in electronegativity, the shorter the bonds as compared with the sum of the covalent radii of the two halogens. Iodine monchloride, ICl, forms zig-zig chains in both its α and β forms, the forms differing in whether the Cl branches are cis (α) or trans (β).

Structure and Shape of AB Tpye Intyerhalogen Compounds

In AB type molecules A halogen atom is less electronegative than B halogen atom. From the Lewis structure of AB molecule. ${}^{\times}_{\times}\overset{\times}{\underset{\times}{A}}{}^{\times}_{\times}\ \times\dot{\ddot{\underset{\cdot\cdot}{B}}}:$ or ${}^{\times}_{\times}\overset{\times}{\underset{\times}{A}} - \dot{\underset{\cdot\cdot}{B}}:$ it is evident that the central atom A (central atom is that which is less electronegative) has been surrounded by one σ–bonding electron pair and three lone pairs of electrons, *i.e.*, σ–bps + lps = 1 + 3 = 4.

Consequently the central atom A is sp^3 hybridised in AB molecule. One can understand the occurrence of sp^3 hybridisation in AB type interhalogen molecules by taking the example of ICl molecule. The sp^3 hybridisation undergone by I-atom has been depicted as follows :

$5s^2$. $5px^2$ $5py^2$ $5px^1$ (I-atom in its ground state) $\xrightarrow{sp^3 \text{ hybridisation}}$ $(sp^3)^2$ $(sp^3)^2$ $(sp^3)^1$ [Four sp^3 hybrid orbitals three of which contain a lone pair of electrons while the fourth hybrid orbital is singly-filled].

sp^3 hybridisation scheme depicted above makes it evident that three of the four sp^3 hybrid orbitals have one lone pair (lp) each while the fourth sp^3 hybrid orbital has one electron. This singly-filled sp^3 hybrid orbital makes a head-to-head (linear) overlap with the singly-filled $3p_z$ orbital on Ci-atom and forms one sp^3 (I)– px(Ci) σ–bond.

Although the spatial arrangement of the four electron pairs (bp = 1 and lps = 3) round I atom is tetrahedral, because of the presence of three lone Paris of electrons in three sp^3 hybrid orbitals, the shape also having linear shape.

STRUCTURE AND SHAPE OF AB_3 TYPE INTERHALOGEN MOLECULES

(Examples : ClF_3, BrF_3, ICl_3), Lewis structure of AB_3 molecule,

$$\begin{array}{c} B \\ | \\ B-\underset{\cdots}{A}-B \end{array}$$ shows that the central atom A is surrounded by three σ-bounding electron pairs and two lone Pairs of electrons, *i.e.*, σ – bps + lps = 3 + 2 = 5. Consequently the central atom, A is sp^3d hybridised in AB_3 molecule. The occurrence of sp^3d hybridisation in AB_3 type interhalogen molecules can be understood by taking the following examples :

(i) Structure and shape of ClF_3 molecule.

In this compound, chlorine is the central atom and three fluorine atoms are arranged around it.

The electronic configuration of fluorine atom is

$${}_9F = 1s^2\ 2s^2\ 2p_x^2\ 2p_y^2\ 2p_z^1$$

The electronic configuration of chlorine atom is

$${}_{17}Cl = 1s^2\ 2s^2\ sp^6\ 3s^2\ 3p^5$$

and that of excited state of chlorine atom is

$$Cl\ (excited) = 1s^2\ 2s^2\ 2p^6\ 3s^2\ 3p^4\ 3d^1$$

Electronic structure of chlorine atom (excited state)

↑↓ ↑↓ ↑ ↑ ↑ ↑

sp^3 hybridisation
Trigonalbipyramid with two positions
occupied by lone pairs

In the excited state of chlorine atom, three unpaired electrons are available which form three bonds with three flurorine atoms involving sp^3d hybridisation. Therefore, ClF_3 may have trigonal bipyramid structures like I, II and III with two positions occupied by lone pairs.

It is not possible to predict which of the three possible arrangements I, II and III will be formed. From micro-wave data, the structure of ClF_3 is planar and T - shaped with bond angles almost 87°40' which is close to 90°. This indicates structure (III).

The repulsion between the two lone pairs of electrons reduces F_aClF_b bond angle from 90° to 87°40' Both F_a – Cl bond lengths are equal (= 1.70 Å) while Cl – C_b bond length (= 1.60 Å) is different from these two bond lenghts. The difference in bond lengths is in accordance to our expectations, since a trigonal bipyramid is not a regular shape. Equatorial bonds (those in the triangle) are different from epical bonds (those pointing up and down).

The X-ray structure of crystalline ClF_3 shows that the molecule ClF_3 is T-shaped with a bond angle of 87° and bond lengths 1.716 A° and 1.621 A°.

(ii) Structure and shape of BrF_3 : Like ClF_3, BrF_3 molecule is also having bent T-shaped structure. In this molecule Br-atom is the central atom and is sp^3d hybridised. Here also, lone pairs occupy the basal positions of the trigonal bipyramid. The structure of BrF_3 (by micro wave) is also T-shaped with a bond angle of 86°12° and bond lengths of 1.810 Å and 1.721Å.

(iii) Structure and shape of ICl_3 molecule. Where as ClF_3 and BrF_3 are monomeric T-shaped molecules, ICl_3 in the solid state is dimeric having a chlorine-bridged structure shown in

The dimer, $(ICl_3)_2$ gets formed by joining two T-shaped ICl_3 molecules and is flat. Each I-atom is sp^3d hybridised and gets bonded to two bridging Cl-atoms and two terminal Cl-atoms. Each I-atom is having two lone pairs of electrons due to which $(ICl_3)_2$ molecule does not have the expected regular trigonal bipyramida shape but is having a square planer shape.

STRUCTURE AD SHAPE OF AB_5 TYPE INTERHALOGEN MOLECULES

(Examples : IF_5, ClF_5 BrF_5). The Lewis structure of AB_5 molecule, Shows that the central atom A is surrounded by five σ-bonding electron pairs and one lone pair of electrons, *i.e.*, σ – bps + lps = 5 + 1 = 6. Consequently the central atom, A is sp^3d^2 hybridised in AB_5 molecule. The occurrence of sp^3d^2 hybridisation in AB_5 type interhalogen molecules can be understood by considering the following examples :

(i) Structure and shape of IF_5. In this molecule, iodine is the central atom in the molecule. The electronic configuration of fluorine atom is

$$_9F = 1s^2\ 2s^2\ 2p_x^2\ 2p_y^2\ 2p_z^1$$

The electronic configuration of iodine atom is given as

$$_{53}I = 1s^2\ 2s^2\ 2p^6\ 3s^2\ 3p^6\ 3d^{10}\ 4s^2\ 4d^{10}\ 5s^2\ 5p^5$$

and that of excited state is pur as :

$_{53}I$ (excited) = $1s^2\ 2s^2 2p^6\ 3s^2\ 3p^6\ 3d^{10}\ 4s^2\ 4d^{10}\ 5s^2\ 5p^3\ 5d^2$

In the excited iodine atom, five unpaired electrons are available, resulting in sp^3d^2 hybridisation Thus, IF_5 is formed by sp^3d^2 hybridisation and has, therefore, an octahedral shape with one position occupied by lone electron pair

Although, the spatial arrangement of six electron pairs (σ–bps = 5 and lp = 1) round the central I-atom is octahedral, due to the presence of one lone pair of electrons in the axial hybrid orbital, the shape of IF_3 gets distorted and becomes square pyramidal.

A recent single crystal structure study of this molecule made by Burbank and James bas shown that the basal F– atoms (F_b atoms) are slightly displaced upwards from the base of the square pyramid and hence IF_5 molecule assumes the structure.. F_aIF_b bond angle has been found to be equal to 81.9°. I – F_a bond length (= 1.73 Å) is shorter than the I – F_b bond length (= 1.87 Å.)

(ii) Structure and shape of BrF_5. Raman measurements on purified liquid bromine pentafluoride at wavelenght 4358 Cm^{-1} excitation exhibit nine frequencies called fundamentals. Three of these frequencies are very strong. From Raman measurements, it was predicted by Burkey and Jones (1951) that BrF_5 should have tetragonal pyramidal structure.

On the other hand, the other special studies indicated trigonal bipyramidal structure. However, the central bromine atom in BrF_5 is surrounded by six electron pairs; five of which are used to form bonds with five fluorine atoms whereas the remaining sixth corner be occupied by a non-bonded lone electron pair which occupies more space than bonded pair, indicating distorted octahedron for BrF_5 molecule.

STRUCTURE AND SHAPE OF ClF_5 MOLECULE

Like IF_5, ClF_5 molecule also has square pyramidal shape which results due to sp^3d^2 hybridisation of Cl and Br atoms repectively.

STRUCTURE AND SHAPE OF LF

Iodine pentafluoride has a pentagonal bipyramid structure with fluorine atoms at the corners and iodine at the centre. It is probable the only known example of a non-transition using three d orbitals for bonding.

Electronic configuration of fluorine atom is

$F_9 = ls^2\ 2s^2\ 2p_x^{\ 2}\ 2p_y^{\ 2}\ 2p_z^{\ 1}$

The electronic configuration of iodine atom is given by

$I_{53} = ls^2\ 2s^2\ 2p^6\ 3s^2\ 3p^6\ 3d^{10}\ 4s^2\ 4p^6\ 4d^{10}\ 5s^2\ 5p^5$

and that of excited state of iodine atom is

I_{53} (excited) = $ls^2\ 2s^2\ 2p^6\ 3s|\ 3p^6\ 3d^{10}\ 4s^2\ 4p^6\ 4d^{10}\ 4s^1\ 5p^3\ 5d^3$

Thus, seven unpaired electrons are available which form bonds with seven fluorine atoms involving sp^3d^3 hybridisation. As IF7 involves sp^3d^3 hybridisation, it ha a pentagonal bipyramidal structure sp^3d^3 hybridisation scheme above makes it evident that each of the seven hybrid orbitals is singly-filled with electrons.

Each of these hybrid orbitals is having a head-to-head (linear) overlap with the singly-filled $2p_x$ orbital on F atom and forms seven I – F σ bonds and IF_7 molecule is having pentagonal bipyramidal shape which has two axial I – F_a bonds and five basal I – F_b bonds. I – F_b bonds are lying in the pentagonal plane and are directed at an angle of 72° to each other while the remaining two I – F_a bonds get inclined at an angle of 90° to the pentagonal plane. The anlge between two I – F_a bonds is equal to 180°. Thus, F_a – I – F_b = 90°, F_b – I – F_a = 180°. Both I – F_a bond lengths which are equal to each other are greater than both I – F_b bond lengths which are again equal to each other.